Hydrocarbon Synthesis from Carbon Monoxide and Hydrogen

Hydrocarbon Synthesis from Carbon Monoxide and Hydrogen

Edwin L. Kugler, EDITOR

Exxon Research and Engineering Company

F. W. Steffgen, EDITOR

U.S. DOE–PETC

Based on a symposium

sponsored by the Division of

Petroleum Chemistry, Inc. at the

175th Meeting of the

American Chemical Society,

Anaheim, California,

March 13–14, 1978.

ADVANCES IN CHEMISTRY SERIES **178**

AMERICAN CHEMICAL SOCIETY

WASHINGTON, D. C.   1979

Library of Congress CIP Data

Hydrocarbon synthesis from carbon monoxide and
hydrogen.
(Advances in chemistry series; 178 ISSN 0065–
2393)

Includes bibliographies and index.

1. Petroleum—Synthetic—Congresses. 2. Carbon
monoxide—Congresses. 3. Hydrogen—Congresses.
I. Kugler, Edwin L., 1945–   . II. Steffgen, F. W.
III. American Chemical Society. Division of Petroleum
Chemistry. IV. Series.

QD1.A355 no. 178 [TP698] 540'.8s [662'.6623]
ISBN 0-8412-0453-5 79-18286 ADCSAJ 178 1–181
1979

Advances in Chemistry Series

M. Joan Comstock, *Series Editor*

FOREWORD

ADVANCES IN CHEMISTRY SERIES was founded in 1949 by the American Chemical Society as an outlet for symposia and collections of data in special areas of topical interest that could not be accommodated in the Society's journals. It provides a medium for symposia that would otherwise be fragmented, their papers distributed among several journals or not published at all. Papers are reviewed critically according to ACS editorial standards and receive the careful attention and processing characteristic of ACS publications. Volumes in the ADVANCES IN CHEMISTRY SERIES maintain the integrity of the symposia on which they are based; however, verbatim reproductions of previously published papers are not accepted. Papers may include reports of research as well as reviews since symposia may embrace both types of presentation.

CONTENTS

PREFACE

The petrochemical industries traditionally have depended upon petroleum as their source of feedstocks. The 1973 Arab oil embargo emphasized the need to develop alternate sources. A resultant resurgence of interest in Fischer–Tropsch chemistry will undoubtedly be bolstered by the increased costs and decreased availability now being projected for petroleum supplies.

Work in the field of catalyzed hydrogenation of carbon monoxide began some forty years ago. At reasonably low temperatures, the nonselective production of many organic compounds from carbon monoxide and hydrogen is thermodynamically feasible. It is this nonselectivity that is the major barrier to applying this type of synthesis to our petrochemical needs. Recent research focuses on the development of new catalyst systems that maximize more desirable products (i.e., low-molecular-weight olefins and alcohols).

Thorough investigations of these important reactions are now possible, using improved analytical techniques. Slight variations in reaction conditions have been found to effect significant changes in product selectivity. Small, judiciously placed additions of salts, alkali, or even sulfur (once believed to be detrimental in even trace amounts) to the metal catalyst can enhance product selectivity. Supported metal catalysts have greater stability than unsupported, and the nature of the support material also affects the reaction.

Investigations into these topics are presented in this volume. Iron, nickel, copper, cobalt, and rhodium are among the metals studied as Fischer–Tropsch catalysts; results are reported over several alloys as well as single-crystal and doped metals. Ruthenium zeolites and even meteoritic iron have been used to catalyze carbon monoxide hydrogenation, and these findings are also included. One chapter discusses the prediction of product distribution using a computer to simulate Fischer–Tropsch chain growth.

June 29, 1979

Kinetics of CO Hydrogenation on Nickel(100)

D. W. GOODMAN, R. D. KELLEY, T. E. MADEY,
and J. T. YATES, JR.

Surface Science Division, National Bureau of Standards,
Washington, DC 20234

A specially designed ultrahigh vacuum system has been used to examine the effect of surface chemical composition on the kinetics of the catalytic methanation reaction. Surface cleanliness is characterized using Auger Electron Spectroscopy (AES) in an ultrahigh vacuum chamber, and reaction kinetics are determined following an in vacuo transfer of the sample to a catalytic reactor contiguous to the AES chamber. Kinetics of CO hydrogenation (H_2:CO ratio of 4:1 and 120 Torr total pressure) over a Ni(100) surface at 450–700 K are compared with those data reported for polycrystalline nickel and high-area-supported nickel catalysts. Very good agreement is observed between both the specific rates and activation energies measured for high-area-supported catalysts and the single crystal Ni(100) surface.

In recent years ultrahigh vacuum methods have been applied to catalytic studies on initially clean metal surfaces having low surface area. In several instances (the hydrogenolysis of cyclopropane over platinum (1) and the catalytic methanation reaction over rhodium (2) and nickel (3)) a link between ultrahigh vacuum methods and conventional catalytic measurements was established. That is, specific reaction rates over low area ($\sim$ 1–10 cm^2) catalyst samples agreed with specific reaction rates for high area samples ($\sim$ 100 m^2/g). These data suggest that low area, well-characterized samples can be used as models for working catalysts in studies of catalytic reaction mechanisms, as well as in studies of the mechanism of catalyst deactivation and poisoning.

In the present work, we are using a specially-designed ultrahigh vacuum system to examine the effect of surface structure and surface chemical composition on the kinetics of the energy-related catalytic methanation reaction ($3H_2 + CO \rightarrow CH_4 + H_2O$). The catalyst sample is a high-purity single crystal of nickel whose surface is cut to expose (100) planes. The surface cleanliness is characterized using Auger Electron Spectroscopy (AES) in an ultrahigh vacuum chamber, and reaction kinetics are determined following an in vacuo transfer of the sample to a catalytic reactor contiguous to the AES chamber

In this account of work in progress, we report that the kinetics of CH_4 production over initially clean Ni(100) are in excellent agreement with previous data for polycrystalline nickel foil and high-area-supported nickel catalysts. Traces of surface impurities such as iron act as poisons, causing a marked lowering of the reaction rate.

Experimental

The ultrahigh vacuum apparatus being used for these studies is illustrated in Figure 1. The single-crystal Ni(100) catalyst sample is spot-welded to two short nickel wires and is heated resistively. The sample is mounted on a retraction bellows and can be translated horizontally to various positions. In position 1 the surface chemical composition is determined using electron-excited AES; in position 2 the front and back of the crystal can be dosed with catalyst poisons or promoters using a molecular-beam dosing array. Both positions 1 and 2 are in the ultrahigh-vacuum analysis and surface-preparation chamber. In position 3 the catalyst is located in a high-pressure ($P \leq 1$ atm) stirred-fllow catalytic reactor.

The high-purity reactant gases are admitted to the reactor as a 4:1 H_2:CO mixture at a total pressure of 120 Torr. The product CH_4 is detected using a gas chromatograph calibrated with a standard mixture. The specific reaction rate at a given catalyst temperature and gas pressure is the turnover number (4), N_{CH_4} (number of CH_4 molecules produced per site per second). N_{CH_4} was determined by an absolute measure of the amount of CH_4 produced during a fixed time (typically 2000 sec) at catalyst temperatures ranging from 450 to 700 K; the Ni(100) atom density (1.62×10^{15} atoms/cm^2) was used for the number of sites per square centimeter.

Prior to each measurement of catalytic reaction rate, the Ni(100) surface was cleaned using an oxidation–reduction cycle. Figure 2a is an AES spectrum of the Ni(100) crystal after heating; a large impurity sulfur peak is evident. After heating in oxygen at 1×10^{-6} Torr at ~ 1400 K, the sulfur disappears but the surface remains oxygen covered (Figure 2b). After heating in hydrogen at 5 Torr at ~ 800 K for several minutes, followed by heating in vacuum, the clean AES spectrum of Figure 2c results. This is the starting point for all of the kinetic measurements. An AES spectrum from the nickel catalyst after the termination of a reaction rate measurement at ~ 700 K is shown in Figure 2d. A small amount

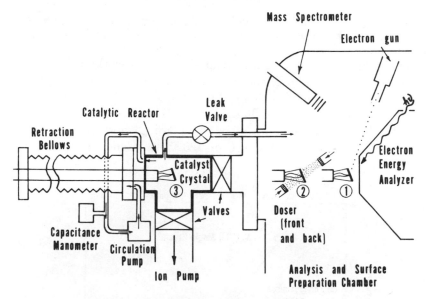

Figure 1. *Ultrahigh vacuum apparatus for studying single-crystal catalysts before and after operation at high pressure in catalytic reactor. Position 1: crystal is in position for Auger-electron-spectroscopy study of surface composition, or for UV photoemission spectrum of surface species. Position 2: crystal is in position for deposition of a known coverage of poisons or promoters for a study of their influence on the rate of a catalytic reaction. Position 3: crystal is in position for a study of catalytic reaction rate at elevated pressures, up to 1 atm. Gas at high pressure may be circulated using pump; mass spectrometric/gas chromatographic analysis of the reactants/products is carried out by sampling the catalytic chamber.*

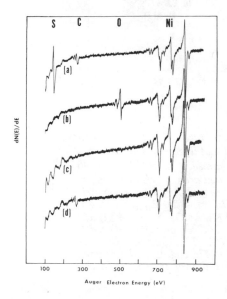

Figure 2. *Auger electron spectra of the Ni(100) catalyst sample under different conditions. (a) AES following repeated heating in vacuum to ~ 1400 K. Impurity sulfur (~ 150 V) and carbon (~ 270 eV) are evident, in addition to the dominant nickel peaks (650–850 eV). (b) AES after heating to ~ 1400 K in oxygen at 1×10^{-6} Torr. Impurity carbon and sulfur are absent, and surface oxygen is present (~ 510 eV). (c) AES after heating surface of (b) at ~ 800 K in hydrogen at 5 Torr. Impurity sulfur, carbon, and oxygen are absent. The broad peaks in the range 100 to 300 eV are believed to be diffraction features (7). (d) AES following methanation reaction in high pressure chamber (4:1 H_2:CO; $P_{total} = 120$ Torr; catalyst temperature ~ 700 K; reaction run for 2000 sec).*

of carbidic-like carbon (5) is evident. Operation at lower temperature (~ 550 K) for a comparable time period results in even lower concentrations of C.

The values of N_{CH_4} determined in the present work are plotted in Arrhenius form in Figure 3; the activation energy determined from the slope of this line is 24.6 kcal/mol. For comparison, the values of N_{CH_4} measured for both polycrystalline nickel foil and high-area-supported nickel catalysts are also shown. The rates are all normalized to a 4:1 H_2:CO mixture at a total pressure of 120 Torr. Generally speaking, a

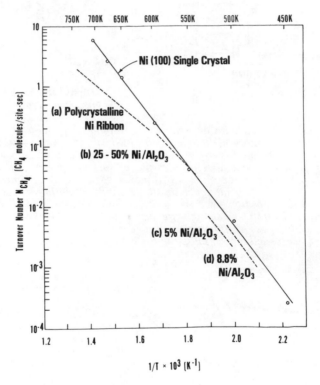

Figure 3. Arrhenius plot of rate of CH_4 synthesis (N_{CH_4}) over different nickel catalysts in various temperature regimes. Ni(100)—present work. $P_{H_2} = 96$ Torr, $P_{CO} = 24$ Torr (geometrical area = 0.85 cm², $E_a = 103$ kJ/mol). (a) Polycrystalline nickel ribbon (geometrical area = 10 cm², $E_a = 66$ kJ/mol (3)) $P_{H_2} = 96$ Torr, $P_{CO} = 24$ Torr. (b) 25–50% Ni/Al_2O_3, $E_a = 84$ kJ/mol (8). (c) 5% Ni/Al_2O_3, $E_a = 105$ kJ/mol (6). (d) 8.8% Ni/Al_2O_3, $E_a = 109$ kJ/mol (6). The data for curves b, c, and d were taken from Table 7b, Ref. 6, and corrected to the hydrogen and CO partial pressures used in this work.

factor of two or three variation in the comparison of turnover numbers is considered good agreement, considering the errors in measurements of reaction rates and active surface area. Thus, there is excellent agreement between both the turnover numbers and activation energies measured for high-area-supported catalysts and single-crystal Ni(100) surfaces. The comparison between values of N_{CH_4} measured on low-surface-area and high-surface-area nickel catalyst samples, as well as the variation in activation energy in Figure 3, have been discussed previously (3). Through the use of the present apparatus, experiments are currently underway on Ni(111) and polycrystalline nickel to investigate the possibility of methanation activity variations on different nickel facets as suggested by the lower turnover numbers measured for the nickel foil compared with the Ni(100) catalyst. It should be pointed out that these earlier data for the polycrystalline nickel were taken in a different apparatus which did not have surface analysis capability.

In several early experiments, severe deactivation of the catalyst sample was observed. AES revealed that large quantities of graphitic-like carbon were present on the deactivated surface, along with small quantities ($\sim$ few tenths of a monolayer) of iron impurity. Qualitatively, it was observed that the carbon and iron scaled with one another. The iron was apparently attributable to traces of impurity iron carbonyls in the reactant CO gas which were efficiently scavenged by the heated nickel catalyst. Storage of the CO over a l-N_2 cooled trap resulted in an iron-free nickel surface following reaction (cf. Figure 2d). In experiments involving kinetic measurements for periods as long as 10^5 sec and total product yield less than 1%, no evidence for self-poisoning and no change from the initial rate was observed.

Conclusions: Future Directions

The present results clearly suggest that well-characterized single-crystal samples can serve as models of practical, working catalysts. The ultrahigh vacuum apparatus described herein will be used further to study the pressure dependence of reaction kinetics, and in particular, the systematics of catalyst poisoning in a quantitative fashion (using the molecular beam doser in conjunction with AES).

Acknowledgment

The authors acknowledge with pleasure the valuable technical assistance of A. Pararas in the design and construction of the high pressure apparatus. This work was supported in part by the U.S. Department of Energy, Division of Basic Energy Sciences.

Literature Cited

1. Kahn, D. R., Petersen, E. E., Somorjai, G. A., *J. Catal.* (1974) **34**, 294.
2. Sexton, B. A., Somorjai, G. A., *J. Catal.* (1977) **46**, 167.
3. Kelley, R. D., Revesz, K., Madey, T. E., Yates, J. T., Jr., *Appl. Surf. Sci.* (1978) **1**, 266.
4. Madey, T. E., Yates, J. T., Jr., Sandstrom, D. R., Voorhoeve, R. J. H., "Treatise on Solid State Chemistry," N. B. Hannay, Ed., Vol. 6B, p. 1, Plenum, New York, 1976.
5. McCarty, J. G., Madix, R. J., *J. Catal.* (1977) **48**, 422.
6. Vannice, M. A., *Catal. Rev.* (1976) **14**, 153.
7. Becker, G. E., Hagstrum, H. D., *J. Vac. Sci. Technol.* (1974) **11**, 284.
8. Bousquet, J. L., Teichner, S. J., *Bull. Soc. Chim. Fr.* (1969) 2963.

RECEIVED June 22, 1978.

Hydrocarbon Synthesis Using Catalysts Formed by Intermetallic Compound Decomposition

A. ELATTAR, W. E. WALLACE, and R. S. CRAIG

Department of Chemistry, University of Pittsburgh, Pittsburgh, PA 15260

Transformed rare earth and actinide intermetallic compounds are shown to be very active as catalysts for the synthesis of hydrocarbons from CO_2 and hydrogen. Transformed $LaNi_5$ and $ThNi_5$ are the most active of the materials studied; they have a turnover number for CH_4 formation of 2.7 and 4.7 $\times$ 10^{-3} sec^{-1} at 205°C, respectively, compared with $\sim$ 1 $\times$ 10^{-3} sec^{-1} for commercial silica-supported nickel catalysts. Nickel intermetallics and $CeFe_2$ show high selectivity for CH_4 formation. $ThFe_5$ shows substantial formation of C_2H_6 (15%) as well as CH_4. The catalysts are transformed extensively during the experiment into transition metal supported on rare earth or actinide oxide. Those mixtures are much more active than supported catalysts formed by conventional wet chemical means.

This chapter is largely concerned with the behavior in regard to heterogeneous catalysis of intermetallic compounds in which one component is an actinide (thorium or uranium) or a rare earth (designated R) and the other is a $3d$ transition metal.

The 14 rare earth elements and chemically similar yttrium and thorium are prolific formers of intermetallic compounds. These compounds exhibit a variety of interesting and unusual physical characteristics—spiral magnetic structures (1,2), crystal field effects (including Van Vleck paramagnetism) (3,4), giant magnetostrictions (5), giant magnetocrystalline anisotropies (6), etc. Structural, magnetic, and thermal properties of these materials have been collated and summarized in a recent monography by Wallace (7).

0-8412-0453-5/79/33-178-007$05.00/0
© 1979 American Chemical Society

Along with attracting interest from a fundamental point of view, the rare earth intermetallics in recent years have begun to attract widespread attention because of the technological implications of some of their observed physical and chemical properties. They are highly regarded as new materials for the production of high energy magnets and as hydrogen storage media. Both of these are of very considerable significance in regard to the national energy issue.

The solvent power of rare earth intermetallics, specifically LaNi₅, was first demonstrated by Neumann (8) and shortly thereafter by Van Vucht, Kiujpers, and Bruning (9). The latter investigators found that hydrogen is absorbed rapidly and reversibly at room temperature. This feature of the rare earth intermetallics is exemplified (10) for ErCo₃ and HoCo₃ in Figure 1. In the experiment summarized by these plots a container pressurized with hydrogen is brought in contact with the intermetallic. It is noted that pressure drops after a few seconds because of hydrogen absorption by the metal, and the solid reaches a saturation concentration in less than two minutes.

Neutron diffraction studies of hydrogenated rare earth intermetallics show (11, 12, 13) that hydrogen is present in the lattice as a monatomic species. This establishes that hydrogen is absorbed dissociatively. The existence of monatomic hydrogen at the surface, if only fleetingly, suggested that this class of alloys warranted attention as hydrogenation catalysts.

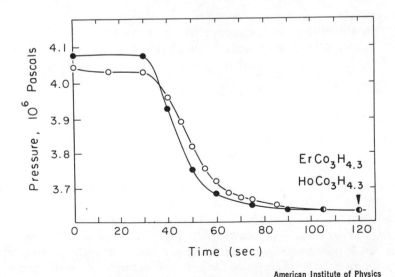

Figure 1.   Absorption of hydrogen by bulk specimens of HoCo₃ and ErCo₃. (●) HoCo₃, (○) ErCo₃.

As is indicated in greater detail below, studies have shown that the intermetallics undergo extensive (perhaps total) decomposition in the course of the reactions that are being investigated, and it is highly probable that it is the decomposition products or the transformed intermetallics which are the active catalysts. Initially, the materials were examined as synthetic ammonia catalysts (*14*). More recently, attention has been directed toward their use as catalysts for the synthesis of hydrocarbons from CO and CO_2.

LaNi$_5$ is the classic example of a rare earth intermetallic with a large capacity for hydrogen. In view of the considerations set forth above, it was one of the earliest numbers of this class of compounds selected for study. Coon et al. (*15, 16*) observed that reaction of CO and H$_2$ over LaNi$_5$ and its decomposition products and the other RNi$_5$ compounds as well began at $\sim 200°C$, and by 380°C 90% of CO was converted in a single pass through the catalyst. Work on the RNi$_5$ class of compounds is reported elsewhere (*15, 16, 17, 18*). More recently attention has turned to intermetallic compounds containing manganese and iron. These studies are summarized in the present chapter, along with some newer results obtained using ThNi$_5$, which appeared in the earlier study to be transformed in the reaction to a substance having unusual activity as a catalyst.

Experimental

The general experimental techniques used have been described in earlier publications from this laboratory (*15, 16, 17, 18*). However, some important modifications have been instituted recently. The reactions were carried out in a fixed bed, single pass differential reactor. Provisions were made to measure in situ the surface area of the fresh and used catalyst. In the previous studies, argon areas were measured by the Nelson–Eggertsen pulse technique (*19*). In the course of the work it was established, as noted above, that a material such as ThNi$_5$ was extensively or totally transformed into ThO$_2$ and Ni (vide infra). It seemed highly probable that the nickel in this instance was the site of the reaction and accordingly there was a need to obtain information about the metallic surface area, or more specifically, the number of active sites, rather than total surface area as is measured with argon absorption. Consequently, chemisorption of CO on the used catalyst also is being measured now. This is accomplished using the pulse technique described by Gruber (*20*) and Freel (*21*). In the present work, helium is used as the carrier gas and CO is used for chemisorption. Since impurities in the rather large amount of carrier gas used, compared with the amount of pulsing gas used, can lead to complications, scrupulous attention was paid to the purity of the helium utilized. He carrier gas of the needed purity was obtained as the evaporate from liquid helium. This proved to be necessary to obtain reproducible chemisorption results.

Pretreatment of samples is traditionally and important aspect of catalytic studies. It is of very minor importance for the materials used in the present work. There appear to be two reasons for this: (1) as is indicated in the following section, the sample is very extensively transformed during the course of the reaction so that pretreatment effects are very rapidly obliterated, and (2) these materials appear to have rather remarkable self-cleaning features, as has been brought out by the very recent UPS work of Siegmann, Schlapbach, and Brundle (22).

Results

Activity of Decomposed ThNi₅. The catalysts formed using the RNi_5 series studied earlier by Coon et al. showed specific activities (based on argon surface areas) larger by one order of magnitude than that of commercial nickel-supported catalysts. The newly obtained chemisorption results lead to the turnover numbers given in Table I. Results for commercially available silica supported nickel catalysts are given for purpose of comparison along with results obtained for transformed $LaNi_5$ and $MmNi_5$ (Mm represents mischmetal). The exceptional activity of tarnsformed $ThNi_5$ and RNi_5 compounds is evident.

As has been alluded to above, the catalysts are extensively transformed when exposed to a mixture of CO and H_2 at $T > \sim 225°C$. This transformation was first noted by Takeshita, Wallace, and Craig (14) in the use of these materials as synthetic ammonia catalysts. RCo_x and RFe_x intermetallics were converted into iron or cobalt rare earth nitride. This was established by conventional x-ray diffraction measurements. Coon (16) observed that the RNi_5 compounds were transformed by the CO/H_2 mixture into R_2O_3 and Ni, and a similar transformation also was observed by Elattar et al. (17) for $ThNi_5$, UNi_5, and $ZrNi_5$. SEM and EDAX results on $ThNi_5$ show the formation of nickel nodules ~ 0.5 μm in diameter situated on a ThO_2 substrate.

From the comments in the preceding paragraph it superficially appears that the work to date consists merely in producing supported catalysts in a new way. The conventional way of producing catalysts involves initially wet chemical procedures followed by a calcination proc-

Table I. Turnover Numbers (N) Measured at 205°C

	$10^3 N\ sec^{-1}$	Ref.
Ni on SiO_2	1.1[a]	23
Ni on SiO_2	0.5 to 1[a]	24
$MmNi_5$	3[b]	25
$LaNi_5$	2.7[b]	This work
$ThNi_5$	4.7[b,c]	This work

[a] Based on hydrogen chemisorption.
[b] Based on CO chemisorption.
[c] $E_{act} = 80.3 \pm 0.5$ kJ/mol.

Table II. Reactivity of CO + 3H$_2$ over Nickel Supported on ThO$_2$^{a}

	W.C. Ni/ThO$_2$	*I.C.D. Ni/ThO$_2$*
Initiation of reaction	245°C	110°C
% CO transformed	1.7 at 510°C	98 at $T \geq 290$°C
% CH$_4$ in effluent gas	Traces at 218°C	7.2 at 209°C

^{a} For designation of W.C. and I.C.D., *see* text. Flow rates and reaction conditions were identical for the two types of catalyst.

ess and then a reduction pretreatment. The new procedure involves, for example, the formation of Ni/ThO$_2$ by decomposition of ThNi$_5$. If these are termed Method W.C. and Method I.C.D. for Wet Chemical and Intermetallic Compound Decomposition, respectively, one might superficially expect identical catalytic features. Experiment fails to confirm this expectation. It is found that I.C.D. Ni/ThO$_2$ has substantially superior catalytic activity to W.C. Ni/ThO$_2$ (Table II), a finding which appears to be of rather considerable significance for Fischer–Tropsch chemistry.

Catalysts formed by intermetallic compound decomposition show impressive resistance to H$_2$S. Results obtained are shown in Figure 2. It is to be noted that decomposed ThNi$_5$ is more resistant to poisoning than decomposed ZrNi$_5$ or the commercial supported nickel catalyst. It is not clear at this time what factors produce these differences. Perhaps the metallic area was smaller for the kieselguhr-supported material; it was not determined. The metallic areas of the two decomposed intermetallics were established and were comparable.

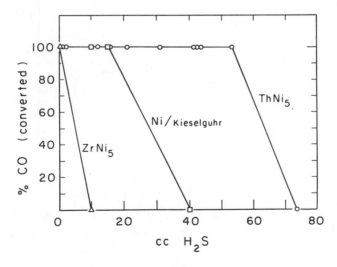

Figure 2. Poisoning of ZrNi$_5$, ThNi$_5$, and Ni/kieselguhr by H$_2$S fed into the feed gas stream. The feed gas had composition 3H$_2$ per 1 CO.

Activity of Decomposed RMn₂, ThFe₅, and CeFe₂. The series RMn_2 (R = Nd, Gd, and Th) has been examined. With these materials only traces of hydrocarbon formation from $3H_2/CO$ mixtures in the temperature range extending to 500°C were observed, even though they are broken down into manganese and R_2O_3. $ThFe_5$ and $CeFe_2$ break down into Fe and ThO_2 or CeO_2. These mixtures are active but the behavior is different from the decomposed nickel compounds in two ways—product distribution and variation of activity with time. These are illustrated in Table III and Figure 3, respectively. As regards product distribution, the nickel intermetallics, which have received the most attention to date, give, if CO_2 and H_2O are excluded, CH_4 almost exclusively. This is also true of $CeFe_2$. However, with $ThFe_5$ significant amounts of C_2H_6 and C_2H_4 were obtained (*see* Table III).

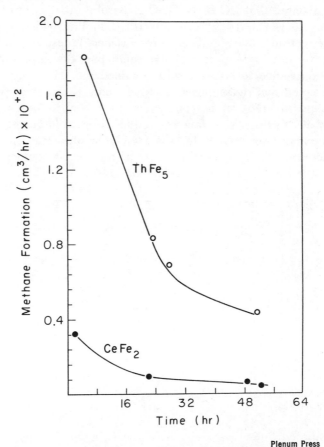

Plenum Press

Figure 3.   Plots showing the declining activity of ThFe₅ and CeFe₂. The ThNi₅ and the RNi₅ catalysts do not show a decrease in activity.

Table III.   Product Distribution[a] **for CO + 3H$_2$ at 285°C**

	Mol % *over ThFe$_5$*[b]	*Mol %* *over CeFe$_2$*
CH$_4$	40	75
C$_2$H$_6$	15	—
C$_2$H$_4$	1	—
CO$_2$	44	25

[a] Exclusive of H$_2$O and deposited carbon.
[b] Traces of C$_3$H$_6$ and C$_3$H$_8$ were also observed.

With CeFe$_2$ and ThFe$_5$ reaction began at $\sim$ 200°C and became maximal at 440°C for CeFe$_2$ (53% of CO converted) and at 460°C for ThFe$_5$ with 77% of the CO converted. These were results in a single pass experiment with a space velocity of 6700 per hour.

There is a dramatic difference between the behavior of CeFe$_2$ and ThFe$_5$ and that of RNi$_5$, ThNi$_5$, and other nickel compounds in that the activity of the iron strongly decreases with time whereas with the nickel compounds activity increased with time. The behavior of the nickel compounds is undoubtedly a consequence of the progressive transformation of ThNi$_5$, UNi$_5$, or RNi$_5$ into nickel plus an oxide, the mixture being the active catalyst. In the case of the iron compounds, the decline in activity is caused by carbon deposition and perhaps the formation of small amounts of waxes since it was observed that the reactor tended to plug up.

Concluding Remarks

The results obtained to date clearly show that very active catalysts can be formed by the decomposition of intermetallic compounds. The two intermetallics ThNi$_5$ and ZrNi$_5$ decompose to give Ni/ThO$_2$ and Ni/ZrO$_2$. The latter is only weakly active and is easily poisoned by H$_2$S. The differing behavior of these two materials has been studied and partially elucidated by Auger spectroscopy and characteristic energy loss spectroscopy (26). It has been shown that nickel in Ni/ZrO$_2$ becomes heavily overlaid with graphite, whereas with thoria as the substrate this does not happen.

Ni/ThO$_2$ produced by conventional wet chemistry means is considerably less active than the same material produced by decomposition of an intermetallic compound. The reason for this difference remains to be elucidated.

Decomposed CeFe$_2$ and ThFe$_5$ are active Fischer–Tropsch catalysts. ThFe$_5$ produces significant amounts of C$_2$ product. The plasmon oscillation behavior of the substrates, which is now under study in this laboratory, exhibits sufficiently different behavior from catalyst-to-catalyst as to suggest a substantially different electronic nature of the oxide substrate and/or the transition metal nodules (27). This probably underlies the variation in catalytic behavior of the several supported nickel catalysts.

Literature Cited

1. Vord, D. G., Givord, F., Lemaire, R., *J. Physiol., Suppl.* (1971) 32, 668.
2. Gallay, J., Hunout, J., Forcinal, G., Deschanvres, A., *C. R. Acad. Sci., Ser. C* (1972) 274, 1166.
3. Tsuchida, T., Wallace, W. E., *J. Chem. Phys.* (1965) 43, 2087, 2885, and 3811.
4. Wallace, W. E., Mader, K. H., *Inorg. Chem.* (1968) 7, 1627.
5. Clark, A. E., *AIP Conf. Proc.* (1974) 18, 1015.
6. Hoffer, G., Strnat, K., *IEEE Trans. Mag.* (1966) 2, 487.
7. Wallace, W. E., "Rare Earth Intermetallics," Academic, New York, 1973.
8. Neumann, H. H., Ph.D. Thesis, Technische Hochschule, Darmstadt (1969).
9. Van Vucht, J. H. N., Kuijpers, F. A., Bruning, H. C. A. M., *Philips Res. Rep.* (1970) 25, 133.
10. Gualtieri, D. M., Narasimhan, K. S. V. L., Takeshita, T., *J. Appl. Phys.* (1976) 47, 3432.
11. Kuijpers, F. A., Loopstra, B. O., *J. Phys. (Paris)* (1971) 32, C 1-658.
12. Andresen, A. F., presented at the International Symposium on Hydrides for Energy Storage, Geilo, Norway, August, 1977. To appear in *International Journal of Hydrogen Energy.*
13. Rhyne, J. J., Sankar, S. G., Wallace, W. E., *Proc. Rare Earth Res. Conf. 13th,* (1978) unpublished data.
14. Takeshita, T., Wallace, W. E., Craig, R. S., *J. Catal.* (1976) 44, 236.
15. Coon, V. T., Takeshita, T., Wallace, W. E., Craig, R. S., *J. Phys. Chem.* (1976) 80, 1878.
16. Coon, V. T., Ph.D. Thesis, University of Pittsburgh (1976).
17. Elattar, A., Takeshita, T., Wallace, W. E., Craig, R. S., *Science* (1977) 196, 1093.
18. Wallace, W. E., Elattar, A., Takeshita, T., Coon, V., Bechman, C. A., Craig, R. S., *Proc. Int. Conf. Electronic Struct. Actinides, 2nd,* (1977) 357.
19. Nelsen, F. M., Eggertsen, F. T., *Anal. Chem.* (1958) 30, 1387.
20. Gruber, H. L., *Anal. Chem.* (1962) 34, 1828.
21. Freel, J., *J. Catal.* (1972) 25, 139.
22. Siegmann, H. C., Schlapbach, L., Brundle, C. R., *Phys. Rev. Lett.* (1978) 40, 972.
23. Takeshita, T., Wallace, W. E., unpublished data.
24. Vannice, M. A., *J. Catal.* (1976) 44, 152.
25. Atkinson, G. B., Nicks, L. J., *J. Catal.* (1977) 46, 417.
26. Moldovan, A. G., Elattar, A., Wallace, W. E., Craig, R. S., *J. Solid State Chem.* (1978) 25, 23.
27. Moldovan, A. G., Wallace, W. E., unpublished data.

RECEIVED September 8, 1978. This work was supported by a grant from the National Science Foundation.

Carbon Monoxide Hydrogenation over Ruthenium Zeolites

P. A. JACOBS, J. VERDONCK, R. NIJS, and J. B. UYTTERHOEVEN

Katholieke Universiteit Leuven, Centrum voor Oppervlaktescheikunde en Colloïdale Scheikunde de Croylaan 42, B-3030 Leuven (Heverlee), Belgium

Carbon monoxide is hydrogenated over ruthenium zeolites in both methanation and Fischer–Tropsch conditions. Ru^{3+} is exchanged in the zeolite as the amine complex. The zeolites used are Linde A, X, Y, and L, natural chabazite, and synthetic mordenite from Norton. The zeolites as a support for ruthenium were compared with alumina. The influence of the nature of the zeolite, the ruthenium metal dispersion and the reaction conditions upon activity and product distribution were investigated. These zeolites are stable methanation catalysts and under the conditions used show a narrow product distribution. The zeolites are less active than other supports. Sintering of ruthenium metal in the zeolite supercages shows only minor effects on methanation activity, although under our Fischer–Tropsch conditions more C_2 and C_3 are formed.

The increasing demand of energy has renewed interest in catalytic production of hydrocarbons from CO and hydrogen. The state of this art has been reviewed recently (*1, 2*). In this respect ruthenium has always been a well-studied metal (*2–8*). Mostly, ruthenium was dispersed on classical supports, such as silica and alumina. At high pressures (1000 atm) high molecular weight polymethylene was built up (*3*), while at moderate pressures (~20 atm.) and high conversion methane was formed with moderate selectivity, CO_2 and C_2^+ hydrocarbons being the other carbon-containing reaction products (*5*). Also at ambient pressure, at 210°C the product from the synthesis reaction over supported ruthenium was found to give not more than 60 mol% of methane, the

0-8412-0453-5/79/33-178-015$05.00/0

other products being composed of higher molecular weight hydrocarbons
(4). Dalla Betta et al., reported (7) that the initial rate of CO hydrogenation was independent of the particle size of supported ruthenium, the
turnover numbers being higher for nickel than for ruthenium. Vannice
(4), on the other hand, found that the turnover numbers for ruthenium
were considerably higher than for nickel.

In this work, experiments at ambient pressure were carried out under
methanation and Fischer–Tropsch conditions. The zeolites as a support
for ruthenium were compared with a more conventional one (alumina).
The influence of the nature of the zeolite, the dispersion of the ruthenium
metal and the reaction conditions upon activity and product distribution
were investigated.

Experimental

Materials. Zeolites A, X, Y, and L were from Union Carbide Corporation, and Zeolite Z was a synthetic large port mordenite from Norton
Company. Chabasite was a crystallographically very pure natural zeolite
from an Hungarian deposit. Zeolite Y* is an aluminum-deficient Y zeolite
prepared by H_4EDTA treatment. The hexammine complex of Ru^{3+} was
from Strem Chemicals. The ruthenium-on-alumina catalyst was from
Ventron.

The ruthenium zeolites were prepared by conventional ion exchange
techniques using the $Ru(NH_3)_6^{3+}$ complex. The complex was decomposed at 300°C under flowing dry helium and the catalyst was reduced
further under hydrogen at different temperatures. The nickel on NaY
zeolite was prepared by conventional ion exchange procedures (9). The
sample was dried and hydrogen reduced at 400°C.

Methods. The CO hydrogenation was carried out in a continuous
flow reactor, operating either in the differential or integral mode. Typical
methanation conditions were: a H_2/CO ratio of 4/1 and GHSV of 3600
hr^{-1}. Typical Fischer–Tropsch conditions were: $H_2/CO = 1/1$ and
GHSV = 1800 hr^{-1}. The reactions in any case were done at atmospheric
pressure. On line to the reactor was attached a Hewlett–Packard model
5830A gas chromatograph equipped with the refinery gas option (UOP
method 539-73). Metal dispersion was measured using temperature
programmed desorption of hydrogen up to 300°C or by chemisorption
of CO at 100°C, assuming linear bonding.

Discrimination between ruthenium metal inside the zeolite pores or
at the external surface was made using a combination of temperature
programmed oxidation and x-ray line broadening (9).

Results and Discussion

Methanation Activity. ACTIVITY AND SELECTIVITY. In Figure 1 are
compared the methanation activity of 0.5 wt % Ru on NaY zeolite and
on alumina, and 1% Ni on NaY zeolite. It is seen that the initial activity
of the two ruthenium catalysts is comparable, while the nickel catalyst is

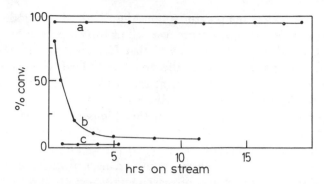

*Figure 1. Methanation activity at 300°C over (a) 0.5%
RuNaY, (b) 0.5% Ru alumina, and (c) 1% NiNaY zeolite*

much less active. Activity maintenance is excellent for the ruthenium
zeolite, while the activity of the particular ruthenium-on-alumina catalyst declines very fast.

The selectivity of 5.6 % Ru on zeolite Y is shown in Figure 2. The
hydrocarbons formed are exclusively methane and only minor amounts
of CO_2 (<6 %). The latter most probably arises from a parallel water
gas shift reaction, since ruthenium metal is known to catalyse this
reaction at an appreciable rate (*10*).

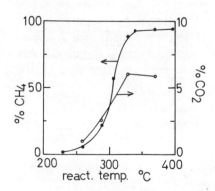

*Figure 2. Selectivity of the metha-
nation reaction over 5.6% RuNaY
zeolite*

INFLUENCE OF THE REDUCTION TEMPERATURE. RuNaY zeolite, containing 5.6 % Ru by weight, is taken as a representative catalyst to
illustrate the influence of the reduction temperature on the methanation
activity. The dispersion of the ruthenium metal phase measured by
desorption of chemisorbed hydrogen and by CO chemisorption is given
in Table I.

For the reduction temperatures considered, all of the ruthenium is reduced to the metallic state. Moreover, all of the metal can be oxidized at 100°C. From this it can be derived that for each catalyst, the ruthenium metal remains dispersed in the zeolite. Indeed, ruthenium metal outside the zeolite cages is oxidized at much higher temperatures (10). It also is seen in Table I that the dispersion measured with hydrogen remains unchanged at increasing reduction temperatures. With CO an optimum value is obtained when the reduction temperature is increased. The CO values can be understood in terms of the following mechanism. Upon decomposition of the ammine complex, part of the ruthenium ions move to the sodalite cages and after reduction remain highly dispersed in these cages inaccessible for CO. At increasing reduction temperatures metal ions migrate from the sodalite to the supercages. This results in an increasing degree of sintering of the metal particles in the supercages. The same mechanism was advanced for platinum zeolites (15). The values from hydrogen chemisorption are not consistent with this mechanism, since hydrogen is able to enter the sodalite cage. They are obtained after desorption at 300°C and therefore may not represent metal surface area variations since either an increased amount of metal dissolved or even spilt over hydrogen may be recovered at increasing reduced temperatures. If hydrogen would not be chemisorbed on monoatomically dispersed ruthenium (in the sodalite cages) (16), this sequence also can be explained.

The influence of the reduction temperature on the turnover number (N—molecules CH_4 formed per second per ruthenium surface atom) is given in Figure 3. The set of values determined with hydrogen gradually decreases. The values obtained with CO show a broad and less pronounced maximum. In view of the above considerations, the chemisorption values with CO better reflect changes in metal surface area of the

Table I. Chemical Characterization of 5.6% RuNaY Reduced at Different Temperatures

Reduction Temperature (°C)	Degree of Reduction	Degree of Reoxidation[a]	H/Ru[b]	CO/Ru[c]
300	100	100	0.46	0.17
350	100	100	0.54	0.33
420	100	100	0.51	0.47
450	100	100	0.46	0.49
480	100	100	0.44	0.35
500	100	100	0.48	0.30

[a] At 100°C.
[b] Amount of hydrogen chemisorbed at 25°C desorbable below 300°C.
[c] Amount CO chemisorbed at 100°C and assuming linear bonding.

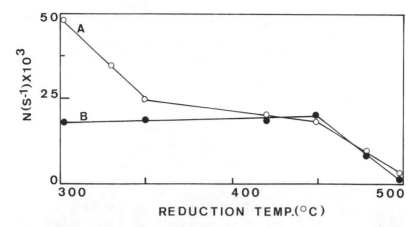

Figure 3.   Influence of the reduction temperature on the turnover number (N) for methane formation at 300°C. (a) N measured with hydrogen desorption at 300°C; (b) N measured with CO chemisorbed at 100°C.

present system. Therefore the N values determined with CO are more realistic. The change of N with the reduction temperatures may be understood as follows. At increasing reduction temperatures, the ruthenium metal particles in the supercages grow at the expense of the metal in the sodalite cages and become more and more encaged. There is hardly any particle size effect on the methanation reaction, just as for ruthenium on alumina (7). At the highest reduction temperatures, steric hindrance most probably causes the decline in N.

Irrespective of the nature of the reaction intermediate, enolic type (11) or surface carbide (12), the decline of the turnover number for the zeolites with higher Si/Al ratio can be explained as follows. For platinum (13) and palladium (14,15) loaded zeolites, support effects are known to exist. The higher the acidity (and the oxidizing power) of the zeolite, the higher will be the electron-deficient character of the supported metal. It also is well established now (16) that the average acidity of hydrogen zeolites increases with the Si/Al ratio. This explains why the electron deficient character of ruthenium should increase with the Si/Al ratio of the zeolite, and a stronger interaction with adsorbed CO should be expected. Vannice (19,20) reported that the N value for CH_4 formation decreases when the heat of adsorption for CO increases. All this explains why the turnover number of the methanation reaction over ruthenium decreases when the Si/Al ratio of the zeolite support increases.

It is expected that the former trend will be less pronounced for higher reduction temperatures, where less chemical interaction between metal and zeolite is expected. The dependence of N upon the Si/Al ratio (Figure 4b) illustrates this.

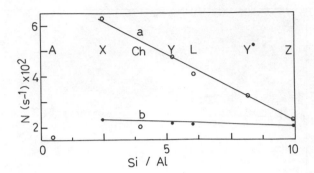

Figure 4. Rate of methanation (N) at 300°C over different zeolites reduced (a) at 300°C or (b) at 500°C

INFLUENCE OF THE NATURE OF THE ZEOLITE. When the Si/Al ratio of structurally different zeolites is varied, N for methane formation also changes (Figure 4), At least, this is true for the small clusters obtained after a 300°C reduction. In the latter case N decreases considerably with increasing Si/Al ratio of the zeolite. A and Ch zeolites do not follow this relation. This is attributable to the inability of these solids to accept the $Ru(NH_3)_6^{3+}$ complex in the inner cages. During the ion exchange procedure this complex is decomposed and probably hydrolyzed ruthenium species are adsorbed. The latter are known to result in much less active catalysts after reduction.

When the ruthenium metal is more sintered (after reduction at 500°C), and possibly strongly encapsulated in the zeolite supercages, the influence of the zeolite on the turnover number for CH_4 formation has almost disappeared.

KINETICS. The kinetic data obtained in a differential reactor over the ruthenium zeolites are consistent with the reaction scheme of Vannice and Ollis (*17*). In this treatment, the rate-limiting step is located in the hydrogenation of an absorbed CHOH species. In the rate expression:

$$r = k' p_{H_2}{}^n p_{CO}{}^{n-(y/2)}$$

y formally is the number of hydrogen atoms involved in the rate-limiting step, and n the power law exponent for the approximation:

$$\frac{K p_{CO} p_{H_2}}{1 + K p_{CO} p_{H_2}} \sim (K p_{CO} p_{H_2})^n$$

In Table II, pertinent kinetic data are shown. Also for RuNaY zeolite a high number for y was found as on alumina. When absolute values are compared, they are considerably lower on zeolite Y. The zeolite structure

Table II.   Kinetic Data for CO Methanation over Ruthenium on NaY Zeolites and on Alumina

	Ru on Zeolite	Ru on Alumina[a]
y	3.84[b]	4.4[c]
$N \times 10^3$ (s^{-1})	48.9	181.0
E_a (kJ/mol)	65.0	76.8

[a] From Ref. 4.
[b] At 300°C.
[c] At 275°C.

compared with alumina seems to inhibit the methanation reaction. Previous paragraphs show that the less acidic zeolites inhibit this reaction less than do the more acidic ones.

Fischer–Tropsch (F.T.) Activity. Carbon monoxide also was hydrogenated over ruthenium zeolites under F. T. conditions: low reaction temperature (>260°C), low CO/H_2 ratios (=1), and longer contact times. The catalysts used were 6.6 and 8.6 wt % Ru on Y and X zeolite, respectively.

PRODUCT DISTRIBUTION. Typical product distributions over RuNaY are given in Table III. Only traces of hydrocarbons with carbon number higher than three were detected. The data further show that: (1) the product selectivity is shifted towards C_1 at increasing reaction temperatures, and (2) the olefin/paraffin ratio also declines at higher reaction temperatures. This is entirely consistent with the behavior of the classical F. T. catalysts when they operate in the same conditions and indicates that olefins are the primary products of a F. T. synthesis.

Table III.   Influence of Reaction Temperature on Product Distribution over RuNaY[a]

Temperature (°C)	% Conv.	C_1/C_2	C_1/C_3	$C_2^=/C_2$	$C_3^=/C_3$	C_1/CO_2
215	1.2	2.0	1.8	0	0.6	2.1
225	2.1	4.8	1.5	0	0.5	2.7
240	3.1	6.3	4.8	0	0.4	4.5

[a] Reduced at 500°C and GHSV = 1800 hr^{-1}.

INFLUENCE OF REDUCTION TEMPERATURE. The influence of the reduction temperature on the selectivity of RuNaY is shown in Table IV. Increasing the degree of sintering results in a gradual decrease of the turnover number for CO disappearance. At the same time the product distribution clearly shifts towards higher hydrocarbons, and more CO_2 is formed. The olefin/paraffin ratio has the tendency to decrease. From the data of Table I, it was already deduced that at increasing reduction

Table IV.   Product Distribution over RuNaX at Different
Reduction Temperatures[a]

Reduction Temperature (°C)	$N \times 10^3$ (sec^{-1})[c]	C_1/C_2[b]	C_1/C_3[b]	$C_3^=/C_3$[b]	C_1/CO_2[b]
300	2.78	4.5	5.6	1.8	1.0
400	2.16	3.0	2.8	1.6	0.8
500	1.16	1.7	1.7	1.3	0.7

[a] Reaction temperature: 250°C, GHSV = 1800 hr^{-1}.
[b] Mole ratios.
[c] N determined using CO chemisorption data.

temperatures, ruthenium particles sinter in the supercages of the zeolite and become more and more encaged. The results therefore seem to indicate that the active site for the formation of non-C_1 products requires the assemblage of more metal surface atoms compared with the site needed for methane formation.

INFLUENCE OF THE ZEOLITE ON THE PRODUCT DISTRIBUTION.   When a less acidic support was used for ruthenium, better activity was found under methanation conditions. Using the same argument, under F. T. conditions a higher selectivity for formation of higher hydrocarbons is expected when a less acidic support is used. In this respect, pertinent data are given over RuX and Y zeolites in Table V. The X zeolite is known to be less acidic than the Y zeolite. There is indeed a definite influence of the zeolite matrix in the indicated direction: higher products are formed over zeolite X.

Table V.   Product Distribution over Ruthenium Zeolites[a]

	x	y
C_1/C_2	4.5	6.3
C_1/C_3	2.8	4.8

[a] At 250°C, after reduction at 300°C.

Conclusions

Ruthenium zeolites are active and stable methanation catalysts. Under the Fischer–Tropsch conditions used here they show a narrow product distribution. When the size of the ruthenium particles enclosed in the zeolite cages is increased, there is hardly any effect found on the methanation activity. Under F. T. conditions a higher amount of C_2 and C_3 products are formed. Zeolites are generally less active than other supports. In the class of zeolite supports, the less acidic zeolites act as promoters of the CO hydrogenation: under methanation conditions the activity is increased; under F. T. conditions, the selectivity is shifted towards higher hydrocarbons. With respect to kinetic behavior and influence of reaction temperature on product distribution, the zeolite behaves in the same way a conventional alumina support does.

Acknowledgment

P. A. Jacobs acknowledges a research position as 'Bevoegdverklaard Navorser' from N.F.W.O. (Belgium). H. Nijs is grateful for a research grant from the Belgian Government (Diensten Wetenschapsbeleid, Programma Afvalstoffen). Financial support from the same institution is also acknowledged.

Literature Cited

1. Vannice, M. A., *Catal. Rev.—Sci. Eng.* (1976) **14**(2), 153.
2. Shah, Y. T., Perrotta, A. J., *Ind. Eng. Chem., Prod. Res. Dev.* (1976) **15**(2), 123.
3. Pichler, H., Meier, H., Gabler, W., Gärtner, R., Kioussis, D., *Brennst.-Chem.* (1968) **48**(9), 22.
4. Vannice, M. A., *J. Catal.* (1975) **37**, 449.
5. Karn, F. S., Schultz, J. F., Anderson, R. B., *Ind. Eng. Chem., Prod. Res. Dev.* (1965) **4**(4), 265.
6. Dalla Betta, R. A., Piken, A. G., Shelef, M., *J. Catal.* (1975) **40**, 173.
7. Ibid. (1974) **35**, 54.
8. Bond, G. C., Turnham, B. D., *J. Catal.* (1976) **45**, 128.
9. Jacobs, P. A., Uytterhoeven, J. B., Beyer, H. K., *J. Chem. Soc. Faraday Trans. 1* (1977) **173**, 7745.
10. Verdonck, J., Jacobs, P. A., Uytterhoeven, J. B., unpublished data.
11. Kölbel, H., Tillmetz, K. D., *J. Catal.* (1974) **34**, 307.
12. Ponec, V., *Catal. Rev.—Sci. Eng.* (1978) **18**(1), 151.
13. Dalla Betta, R. A., Boudart, M., *Catal. Proc. Int. Congr. 5th, 1972,* (1973) **2**, 1329.
14. Figueras, F., Gomez, R., Primet, M., ADV. CHEM. SER. (1973) **121**, 480.
15. Gallezot, P., Datka, J., Massardier, J., Primet, M., Imelik, B., *Proc. Int. Congr. Catal. 6th, 1976* (1977) **2**, 696.
16. Jacobs, P. A., "Carboniogenic Activity of Zeolites," p. 57, Elsevier, Amsterdam, Oxford, New York, 1977.
17. Vannice, M. A., Ollis, D. F., *J. Catal.* (1975) **38**, 514.
18. Gallezot, P., Alarcan-Diaz,.A., Dalmon, J. A., Renouprez, A. J., Imelik, B., *J. Catal.* (1975) **39**, 334.
19. Vannice, M. A., ADV. CHEM. SER. (1977) **163**, 15.
20. Vannice, M. A., *Catal. Rev.—Sci. Eng.* (1976) **14**(2), 153.

RECEIVED August 14, 1978.

Carbon Monoxide Hydrogenation over
Well-Characterized Ruthenium–Iron Alloys

M. A. VANNICE

Department of Chemical Engineering, Pennsylvania State University,
University Park, PA 16802

Y. L. LAM[1] and R. L. GARTEN[2]

Exxon Research and Engineering Co., P.O. Box 45, Linden, NJ 07036

The behavior of the CO/H₂ synthesis reaction has been studied over silica-supported Ru–Fe catalysts, and an optimum range in the Ru:Fe ratio was found to exist in which olefin production was maximized and methane formation was minimized. The catalyst samples were characterized by hydrogen and CO chemisorption, x-ray diffraction measurements, and Mössbauer spectroscopy. Alloy formation was verified at different Ru:Fe ratios, and changes in specific activity and selectivity were observed as this ratio varied. Between Ru:Fe ratios of 1/2 to 2, 45 mol% of the total hydrocarbon product was C_2–C_5 olefins while less than 40 mol% was comprised of methane.

A t relatively low temperatures, it is thermodynamically possible to produce many organic compounds from CO and hydrogen. This situation leads to the major problem area in CO/H₂ synthesis reactions—product selectivity. One of the major research objectives today in this area is the development of new catalyst systems which maximize the more desirable products such as low molecular weight olefins and alco-

[1] Current address: Seçao de Quimica, Instituto Militar de Engenharia, Pca Gen. Tiburcío, URCA, ZC82, 20,000 Rio de Janeiro, Brazil.
[2] Current address: Catalytica Associates, Inc., 3255 Scott Blvd., Suite 7-E, Santa Clara, CA 95050.

hols. Alloy catalysts have shown remarkable specificity in reforming reactions (1), and their use in the Fischer–Tropsch reaction might also lead to improvements in selectivity. Supported metal catalysts are of more interest than unsupported metals because of their higher dispersion and greater stability; however, alloy formation is more difficult to demonstrate in these highly dispersed catalyst systems. One of the newer techniques to be applied to the characterization of supported alloy catalysts is Mössbauer spectroscopy. Since ^{57}Fe is one of the most convenient Mössbauer isotopes to study, the application of this technique to the investigation of Fischer–Tropsch catalysts in conjunction with catalytic studies appears to be a promising approach for attempting to relate catalytic properties to the chemical state of the catalyst. Iron and ruthenium are two of the most active catalysts for CO hydrogenation, yet Ru–Fe alloys have not been studied as Fischer–Tropsch catalysts. Therefore, this study was conducted to determine the catalytic behavior of a series of silica-supported Ru–Fe alloys which were characterized by chemisorption measurements, x-ray diffraction, and Mössbauer spectroscopy, with the latter technique being used to demonstrate alloy formation.

Experimental

The experimental apparatus and procedure have been described in detail elsewhere (2, 3). Only a brief description is given in this section. The chemisorption measurements were conducted at room temperature in a mercury-free glass adsorption system capable of achieving a dynamic vacuum of ca. 3×10^{-7} Torr. Pressures were obtained under differential reaction conditions using a microreactor operating at a total pressure of 103 kPa (1 atm). The flow rate of the H_2 + CO feed gas (H_2:CO = 3) was 20 cm^3 min^{-1}, which gave space velocities of 2400–12,000 hr^{-1} since catalyst samples typically ranged from 0.1 to 0.5 g. Total conversions were usually < 5% although maximum conversions of 11% were allowed to occur over catalysts A_l and B_h. Product analyses were obtained on a Hewlett–Packard 7620 gas chromatograph using Chromosorb 102 columns and subambient temperature programming. The Mössbauer spectra were recorded on an Austin Science Associates Mössbauer spectrometer using a ^{57}CO/Cr source. All isomer shifts (δ) and spectra, however, are reported relative to α-Fe. The source drive was slaved to an asymmetric waveform (flyback mode) so that the source was linearly accelerated through the desired velocity range while the data were accumulated in an ND2200 multichannel analyzer.

Unless otherwise stated, all catalyst samples were subjected to a standard pretreatment prior to adsorption or kinetic studies. The pretreatment consisted of heating to 723 K under 50 cm^3 min^{-1} flowing hydrogen and reducing at this temperature for 1 hr. Space velocities varied from 6000 to 30,000 hr^{-1} depending on catalyst sample size. For hydrogen or CO uptake measurements, the samples were then evacuated at 698 K for 1 hr before cooling under dynamic vacuum to 298 K. For kinetic studies, the samples were cooled under flowing hydrogen to the desired temperature. The zero-pressure intercept of the hydrogen isotherm was chosen to represent monolayer hydrogen coverage, while for CO the dual isotherm method was used with the difference at 100 Torr representing irreversible CO adsorption (3). The gases and their purification have been described earlier (3).

The catalysts were prepared by an incipient wetness impregnation technique, using aqueous solutions of $RuCl_3$ and $Fe(NO_3)_3$. Solutions of predetermined concentrations were added dropwise to silica (HS-5 Cab-O-Sil from Cabot Corp.) with constant mixing. Sequential impregnations of ruthenium and then iron were used for the bimetallic catalysts, and all samples were dried overnight at 383 K after any impregnation step. The iron nitrate solution was 93% isotopically enriched in the ^{57}Fe Mössbauer isotope.

Results and Discussion

Two series of supported Ru–Fe catalysts were prepared: one contained a low metals loading of 1–2 wt% designated by a subscript l, and the other contained a higher metals loading of 5–6 wt% designated by a subscript h. Table I represents the composition and adsorption data for the series with high metals loading, while Table II represents the same information for the series with low metals loading. The same trend occurs in both series—hydrogen and CO chemisorption decrease as the fraction of iron in the samples is increased. Assuming that hydrogen or CO adsorbs on both iron and ruthenium surface atoms, a decrease in percent metal exposed occurs as the fraction of iron in the Ru–Fe catalysts is increased. This trend was verified by x-ray line-broadening measurements. The H:M ratios on the fresh samples indicate that the percent metal exposed varied from nearly 50% down to 1%.

Mössbauer studies of all the iron and Ru–Fe/SiO$_2$ catalysts listed in Tables I and II were performed. The samples all exhibited general characteristics which can be amply illustrated by a discussion of the

Table I. Chemisorption on Catalysts with High Metal Loading

Catalyst	Ru (Atom %)	H:M for Fresh Samples	CO:M for Used Samples
A$_h$—5% Ru/SiO$_2$	100	0.44	0.11
B$_h$—5% Ru, 0.1% Fe/SiO$_2$	96.5	0.24	0.13
C$_h$—5% Ru, 1.5% Fe/SiO$_2$	64.8	0.11	0.054
D$_h$—5% Fe/SiO$_2$ (16 hr)	0	~0.012	0.0031

Table II. Chemisorption on Catalysts with Low
Ruthenium–Iron Loadings

Catalyst	Ru (Atom %)	H:M for Fresh Samples	CO:M for Used Samples
A₁—1% Ru/SiO₂	100	0.40	0.15
B₁—1% Ru, 0.1% Fe/SiO₂	84.7	0.26	0.12
C₁—1% Ru, 0.3% Fe/SiO₂	64.8	0.18	0.10
D₁—1% Ru, 1% Fe/SiO₂	35.6	0.081	0.044
E₁—0.5% Fe/SiO₂ (16 hr)	0	—	0.020

results for only a few samples. For Fe/SiO_2 catalysts, Mössbauer spectra of samples containing low iron concentrations (~ 0.1 wt%) showed that the iron could not be reduced below the Fe^{2+} state even after reduction in dihydrogen at 773 K (4). The same reduction treatment of higher iron concentrations (~ 0.5 wt%) produced ferromagnetic iron metal in addition to Fe^{2+}. As a function of iron loading, the samples behaved as though there were a certain number of sites on the silica surface capable of reacting with Fe^{2+} and preventing its reduction to the metal. When these sites were saturated, excess iron could be reduced to the metallic state. This behavior also has been reported for Fe/Al_2O_3 (5).

The addition of ruthenium to Fe/SiO_2 samples led to different chemical states and chemical behavior of the iron. This is illustrated in Figure 1 (A–E) for 0.1% Fe, 5% Ru/SiO_2. The freshly prepared sample contained only ferric ions ($\delta = 0.36$ mm sec⁻¹, $\Delta = 0.86$ mm sec⁻¹ (Δ is the quadrupole splitting)). Exposure of the sample to dihydrogen at room temperature reduced the iron to Fe^{2+} ($\delta = 1.15$ mm sec⁻¹, $\Delta = 2.07$ mm sec⁻¹) as shown in Figure 1B. Such reduction did not occur in the absence of ruthenium. Ruthenium thus catalyzed the reduction of iron at 298 K and exhibited properties similar to those reported and discussed in detail for other iron-noble metal combinations (5, 6).

Reduction of the 0.1% Fe, 5% Ru/SiO_2 sample at 773 K gave Figure 1C, which is markedly different from the Fe^{2+} spectrum obtained in the absence of ruthenium. Assuming the behavior of the $RuFe/SiO_2$ system parallels that for other iron-noble metal combinations, it is reasonable that the spectrum in Figure 1C is attributable to Ru–Fe bimetallic clusters. This question is now considered. When computer analyzed as two peaks, Figure 1C gave an asymmetric doublet with $\delta = 0.38$ mm sec⁻¹ and $\Delta = 0.73$ mm sec⁻¹. The isomer shift (δ) for bulk Ru–Fe alloys, however, is much different. For Ru–Fe bulk alloys, the isomer shift varies slightly with composition from a value of 0.06 mm sec⁻¹ (with respect to α-Fe) for iron as a dilute impurity (<1 atom %) in ruthenium (7) to 0 mm sec⁻¹ for 70–90% Fe in ruthenium (8, 9, 10). The isomer shift comparison and the large asymmetry in Figure 1C

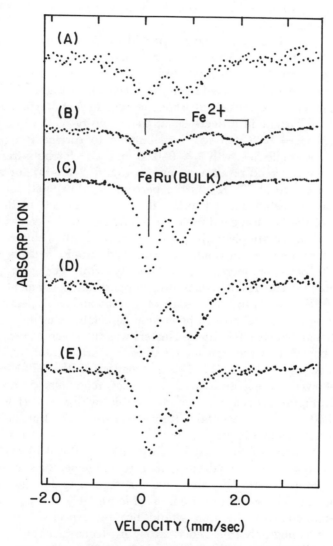

Figure 1. Mössbauer spectra of 0.1% Fe, 5% Ru/SiO₂. (A) As prepared; (B) hydrogen, 298 K; (C) hydrogen, 3 hr, 773 K; (D) evacuated and exposed to oxygen at 298 K; (E) evacuated and exposed to hydrogen at 298 K. All spectra taken sequentially on a single sample at 298 K in 1 atm of treat gas.

suggests an alternate interpretation of the spectrum involving three peaks; a single peak with $\delta = 0$–0.06 mm sec^{-1} attributable to bulk-like iron inside the Ru–Fe clusters and a quadrupole doublet with $\delta \cong 0.38$ mm sec^{-1} and $\Delta \cong 0.73$ mm sec^{-1} attributable to a second kind of iron. Since the percent metal exposed for the sample which gave Figure 1C was 24, it is reasonable to assign the second kind of iron to iron at the surface of Ru–Fe clusters. This is consistent with the presence of a quadrupole doublet because of the asymmetry expected for iron atoms at a surface and with the general experience in Mössbauer studies of bimetallic catalysts. Additional experimental studies also supported an assignment involving surface and bulk peaks for the spectrum in Figure 1C. Thus, if the assignment is correct, decreasing the percent metal exposed of the samples should result in a decrease in the intensity of the quadrupole doublet with $\delta = 0.38$ mm sec^{-1} attributable to surface iron and an increase in the single line with $\delta = 0$–0.06 mm sec^{-1} for bulk iron. This is, indeed, the observed result as shown in Figure 2 (A–D) for a series of samples reduced at 773 K. As the percent metal exposed for Ru–Fe samples was decreased from 54 to less than 1, the intensity of the right-hand peak of the quadrupole doublet at ~ 0.75 mm sec^{-1} decreased until it was absent in the sample with $< 1\%$ metal exposed. The intensity of the peak near 0 mm sec^{-1} attributed to iron inside Ru–Fe clusters showed the opposite behavior, increasing in intensity with decreasing percent metal exposed. The results, therefore, support an assignment of the spectrum in Figure 1C to the overlap of surface and bulk peaks attributable to iron associated with ruthenium as bimetallic clusters.

Additional evidence for Ru–Fe clusters was the reversible oxidation–reduction behavior of the iron in the silica-supported catalysts. This is demonstrated in Figure 1 (C–E). Evacuation of the 773-K-reduced sample followed by exposure to dioxygen at room temperature gave Figure 1D. Figure 1D is attributed to a doublet with $\delta = 0.41$ mm sec^{-1} and $\Delta = 1.10$ mm sec^{-1} attributable to oxidized iron (Fe^{3+}) at the surface of Ru–Fe clusters and a single peak with $\delta = 0$–0.06 mm sec^{-1} attributable to iron in the interior of the clusters. Evacuation of this oxidized sample followed by treatment in dihydrogen, all at room temperature, re-reduced the surface iron (Figure 1E) and gave a spectrum essentially identical to that of the 773-K-reduced sample (Figure 1C). This reversible oxidation–reduction of the iron would not be expected for isolated particles containing only iron and is strong evidence that the ruthenium and iron are associated in bimetallic clusters. The room temperature, reversible oxidation–reduction appears to be characteristic of iron in bimetallic clusters with noble metals. This aspect of the chemical behavior of iron-noble metal catalysts has been discussed in detail by Bartholomew and Boudart (11) and by Garten and Ollis (5).

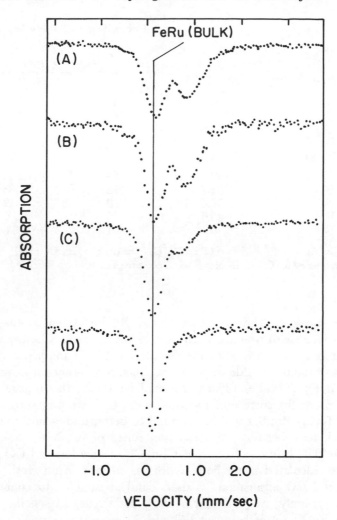

Figure 2. Mössbauer spectra of Ru–Fe samples of different percent-metal exposed, P. All samples reduced 3 hr in hydrogen at 773 K and spectra taken at 298 K in 1 atm hydrogen. (A) 0.1% Fe, 1% Ru/SiO₂, P = 54; (B) 0.1% Fe, 5% Ru/SiO₂, P = 28; (C) 0.1% Fe, 10% Ru/SiO₂, P = 4; (D) Ru–Fe powder (15 atom % Fe), P = < 1.

In summary, the behavior of the RuFe/SiO₂ system, as revealed by Mössbauer spectroscopic studies, parallels that reported for a number of supported bimetallic catalysts containing iron. The isomer shifts for bulk-like iron in the clusters, the trends in the Mössbauer spectra with the percent metal exposed, and the reversible oxidation–reduction behavior of the iron at room temperature are all consistent with Ru–Fe

Table III. CO Hydrogenation

Catalyst	Ru (Atom %)	N_{CH_4} $(sec^{-1} \times 10^3)$	
		a	b
A_h	100	68	270
B_h	96.5	217	402
C_h	64.8	3.5	7.0
D_h	0	16	156
15% Fe/Al$_2$O$_3$	0	—	57
A_1	100	88	234
B_1	84.7	18	38
C_1	64.8	19	32
D_1	35.6	5	9
E_1	0	1	6

[a] $N_{CH_4} = Ae^{-E_{CH_4}/RT} P_{H_2}^x P_{CO}^y$ ($T = 275°C$, $P = 103$ kPa, $H_2:CO = 3$); N_{CO} is the turnover frequency for CO molecules reacted per site per sec.

bimetallic clusters. The exception in the Ru–Fe system is the isomer shift associated with iron at the surface of the clusters. The isomer shift for the iron at the surface of the Ru–Fe clusters is markedly different from that of the iron inside the clusters. Such a difference was not found for the Pd–Fe or Pt–Fe systems. An explanation for the unusual isomer shift of iron at the surface of the Ru–Fe clusters is not apparent at this time and further studies will be required to better understand this result.

The kinetic behavior of these two series of catalysts is shown in Table III. The turnover frequencies for CH$_4$ formation and CO conversion were calculated using both hydrogen adsorption on fresh catalyst samples and CO adsorption on used catalyst samples to count active sites. Surprisingly, we find that a marked decrease in specific activity occurs at Ru:Fe ratios between 1/2 and 2, regardless of one's choice of adsorbate to count active sites. However, the most interesting, and the most important, result is the fact that the N_{CH_4} value is decreased much more drastically than the N_{CO} value. In other words, the methanation activity is inhibited much more than the overall CO hydrogenation activity to form higher molecular weight hydrocarbons. A noticeable change occurs in the activation energy for methane formation when this inhibition occurs. Concurrently, the CO partial pressure dependence shifts from near − 1 toward zero order, but the hydrogen pressure dependence remains close to first order for all alloy compositions. In general, the reduction in the N_{CH_4}/N_{CO} ratio is paralleled by changes in the kinetic parameters in the power–law rate expression for methane formation.

over Ruthenium–Iron Catalysts[a]

N_{CO} $(sec^{-1} \times 10^3)$		E_{CH_4} (kcal/g-mol)	X	Y
a	b			
90	360	27	1.1	−0.8
> 110	> 220	29	1.3	−1.2
22	43	17	1.2	−0.2
77	767	18	—	—
—	160	21	1.1	−0.1
101	270	27	1.2	−0.9
35	77	24	1.3	−0.9
42	73	23	1.3	−0.5
18	33	8	0.8	−0.1
2	12	9	—	—

[b] Based on $H_{(ad)}$ on fresh sample.
[c] Based on $CO_{(ad)}$ on used sample.

Finally, we mentioned at the beginning that alterations in product selectivity represent our major interest. Table IV shows product distributions for both series of catalysts. As expected, the selectivity shift toward higher molecular weight products with increasing iron content is apparent, with methane formation dropping to 40 mol% in some cases. However, we find not only an enhancement in selectivity to heavier hydrocarbons, but also a pronounced increase in the olefin:paraffin ratios when alloy compositions with Ru:Fe ratios between 1/2 and 2 are formed. The production of olefins is enhanced compared with either ruthenium-only or iron-only catalysts. In the case of catalyst C_h in Table V, 45 mol%

Table IV. Hydrocarbon Selectivities over Ruthenium–Iron Catalysts (Mol %)[a]

Catalyst	Ru (Atom %)	C_1	C_2 Ole.	C_2 Par.	C_3 Ole.	C_3 Par.	C_4 Ole.	C_4 Par.	C_5^+
A_h	100	72	tr	9	2	3	tr	7	8
B_h	96.5	73	0.3	10	2	5	0.6	4	5
C_h	64.8	40	11	8	20	tr	9	2	10
D_h	0	59	3	12	15		1.5	6.5	4
A_1	100	92	0	6	0	2	0	1	tr
B_1	84.7	77	2	9	6	1	1	2	2
C_1	64.8	72	2	10	6	3	1	3	3
D_1	35.6	54	9	10	14	tr	5	2	6
E_1	0	69	9	12	10		0	tr	0

[a] $P = 103$ kPa, $H_2:CO = 3$, $T = 250$–$255°C$.

of the total hydrocarbon product is C_2–C_5 olefins. Similar behavior is observed with unsupported Ru–Fe alloys, with methane production dropping below 30 mol% in some cases (12). The optimum set of operating variables, such as percent conversion, pressure, and H_2/CO ratio, required to maximize the olefin:paraffin ratio and to minimize methane production has not yet been determined.

It is interesting to note that the addition of ruthenium to iron produces effects on selectivity similar to those observed when alkali promoters are added to iron; that is, methane production is reduced and olefin formation is enhanced. However, alkali metals also tend to enhance the formation of oxygenated compounds which results in a less favorable selectivity and in product separation problems. Under our reaction conditions, we see no evidence for the formation of any oxygenated compounds. It will be interesting to see if this olefin:paraffin ratio over Ru–Fe catalysts can be enhanced at higher pressures, as is presently achieved with typical, promoted Fischer–Tropsch catalysts, without the concomitant production of less desirable oxygenated compounds. It appears from this study that alloy catalysts can provide favorable shifts in selectivity, and future studies should provide further evidence of this capability.

Acknowledgment

This study was conducted at the Corporate Research Labs, Exxon Research and Engineering Co., Linden, New Jersey. We would like to thank Donna Piano and Larissa Tureaw for performing much of the experimental work.

Literature Cited

1. Sinfelt, J. H., *Science* (1977) **195**, 641.
2. Vannice, M. A., *J. Catal.* (1975) **37**, 449.
3. Vannice, M. A., Garten, R. L., *J. Mol. Catal.* (1975/76) **1**, 201.
4. Garten, R. L., "Mössbauer Effect Methodology," Vol. 10, p. 69, Plenum, New York, 1976.
5. Garten, R. L., Ollis, D. F., *J. Catal.* (1974) **35**, 232.
6. Garten, R. L., *J. Catal.* (1976) **43**, 18.
7. Wortmann, G., Williamson, D. L., *Hyperfine Interact.* (1975) **1**, 167.
8. Ohno, H., Mekata, M., Takake, H., *J. Phys. Soc. Jpn.* (1968) **25**, 283.
9. Wertheim, G. K., Jaccarino, V., Wernick, J. H., Buchanan, D. N. E., *Phys. Rev. Lett.* (1964) **12**, 24.
10. Gupta, S., Lal, K. B., Scrinivasan, T. M., Rao, G. N., *Phys. Status Solidi A* (1964) **22**, 707.
11. Bartholomew, C. H., Boudart, M., *J. Catal.* (1973) **29**, 278.
12. Vannice, M. A., Garten, R. L., unpublished data.

RECEIVED June 22, 1978.

Kinetics of CO + H$_2$ Reaction over Co–Cu–Al$_2$O$_3$ Catalyst

CHEN-HSYONG YANG, F. E. MASSOTH, and A. G. OBLAD

Department of Mining and Fuels Engineering, University of Utah, Salt Lake City, UT 84112

Coprecipitated catalysts containing Co–Cu–Al$_2$O$_3$ have been shown to give improved selectivity to light hydrocarbons for the CO + H$_2$ reaction. This chapter deals with the effect of process variables on catalytic activities and selectivities for one catalyst composition. Runs were carried out in a fixed-bed reactor. The ranges of variables studied were: 225°–275°C, 10–750 psig, 1/1–3/1 H$_2$/CO, and 0.1–1.0 cm^3(STP)/sec g catalyst. Conversion of CO was found to increase with pressure, H$_2$/CO ratio, and temperature. Major hydrocarbon products were: C$_2$–C$_4$ > C$_1$ > C$_5^+$ > CH$_3$OH. The desired C$_2$–C$_4$ fraction increased with H$_2$/CO ratio and decreased with temperature. Water was the predominant, nonhydrocarbon product of the reaction. The reaction rate was found to be first order in hydrogen and inverse square-root order in CO, with an activation energy of 24 kcal/mol. No promoting effect of potassium or sodium was observed.

In a previous paper (1), we have reported on the conversions and selectivities of various Co–Cu–Al$_2$O$_3$ catalysts for the reaction of CO and H$_2$ to light hydrocarbon products. An interesting synergism was found in this catalyst system with respect to metal dispersions and catalyst selectivities. Thus, comparison of CoCu catalyst with that of the separate metal catalysts showed: (1) a significant increase in metal area dispersion and (2) a drastic change in product selectivity, with an increase in C$_2$–C$_4$ products and a decrease in CH$_4$ and CH$_3$OH. The best formulation for maximum C$_2$–C$_4$ selectivity was about 0.7 wt fraction copper. Figure 1

0-8412-0453-5/79/33-178-035$05.00/0

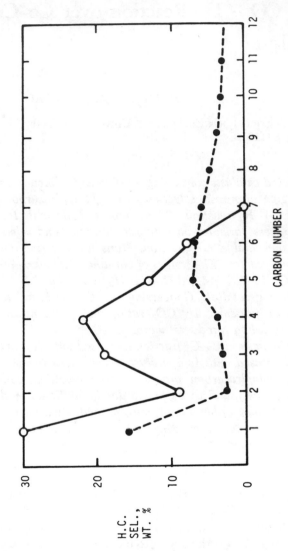

Figure 1. Comparison of molecular weight distribution of hydrocarbon products. (– – –) Co/ThO$_2$, 200°C, 7 atm, O/P = 0.25 (2). (——) CoCu/Al$_2$O$_3$, 250°C, 35 atm, O/P = 0.82.

compares product distributions for the catalyst used in the present study with that of a cobalt–thoria catalyst (2) and illustrates the enhanced C_2–C_4 selectivity achievable.

Since selectivities of this catalyst system were appreciably different from those obtained by Fischer–Tropsch catalysts on the one hand and by methanation catalysts on the other, it seemed desirable to study this interesting system in more detail.

Experimental

Catalysts were prepared by a coprecipitation method described previously (1). The catalyst used in the processing and kinetic studies was prepared by slow addition of a solution of the nitrates of cobalt, copper, and aluminum to a solution of Na_2CO_3. Analysis gave: 5.0% Co, 10.4% Cu, and 0.5% Na on an oven-dried basis.

To test the effect of added potassium, another catalyst was prepared using $(NH_4)_2$ CO_3 in place of Na_2CO_3. A portion of the oven-dried catalyst was then impregnated with KNO_3, followed by a second oven-drying.

Catalysts were pretreated in the reactor by passing a stream of H_2 over at 5 psig for two hours at 250°C followed by four hours at 510°C. The reactor testing procedure and analysis were identical to that reported before (1). Run conditions covered were: 225°–275°C, 10–750 psig, 1/1–3/1 H_2/CO and 0.1–1.0 cm ^{3}(STP)/sec g catalyst.

Results and Discussion

Catalyst Aging. Since some catalyst instability was observed in preliminary tests, a standard condition was chosen which was repeated throughout a series of runs. Two series of runs were conducted on two batches of catalyst. Figure 2 shows catalyst variations experienced during these runs. The values pertain to the standard condition only. Between bars connecting the points, run conditions were varied. An average standard conversion for each day was then used to correct conversions to the first-day base condition. This is similar to the bracketing technique described by Sinfelt (3). After ending with the standard run condition, hydrogen was flowed over the catalyst overnight.

Some variation in CO conversion between days is evidenced in Figure 2. Activity seemed to increase and then decrease over the run period. The catalyst in series B started out slightly more active than that of series A. The two nonhydrocarbon products of reaction are H_2O and CO_2. The high ratio of H_2O/CO_2 of 4:8 indicates that the preferred reaction product is H_2O. This appeared to increase in series A but remained essentially constant during series B. The H_2O values were

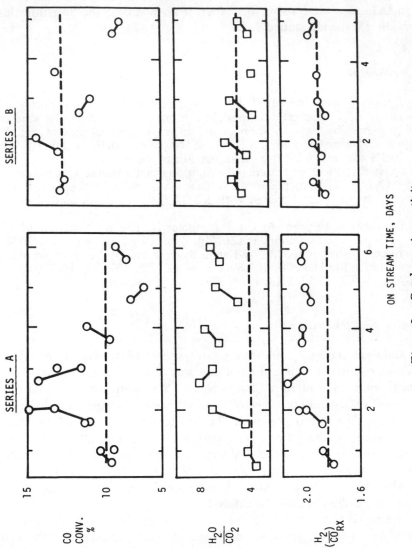

Figure 2. Catalyst aging activity

calculated by a material balance while the CO_2 was directly measured. A similar response is observed in the H_2 to CO reacted since high H_2 consumption is largely manifested in high H_2O yields.

Effect of Process Variables. The effects of space velocity, pressure, and inlet H_2/CO ratio (R) on CO conversion are given in Figure 3. Conversion increased with increase in space time, pressure and H_2/CO ratio, as expected. These runs were all conducted at 250°C. Series A was run at a constant H_2/CO of 7/3 and series B at 500 psig.

The effects of these parameters were qualitatively assessed on catalyst selectivity. Table I summarizes these in terms of the direction and approximate extent of the response. Quantitative values of the responses could not be given because of the data scatter and the general nonlinear nature of the responses over the parameter ranges used. Hydrocarbon products were grouped into four categories: C_1, C_2–C_4, C_5^+, and ROH. The latter was mostly methanol. In addition, the CO_2 yield and olefin-to-paraffin ratio (O/P) are also included. Longer space times increased conversion and CO_2 yield while decreasing H_2O/CO_2 ratio and O/P ratio, but had no effect on hydrocarbon selectivities. Apparently, all processes are increased proportionally so that chain growth remains constant. Increase in the H_2/CO ratio decreased CO_2 and O/P and increased C_1 at the expense of C_2–C_4 and C_5^+. Pressure had negligible effect on selectivities. Higher temperatures favored CO_2, C_1, and C_5^+ with lower C_2–C_4 and ROH.

To maximize C_2–C_4 with this catalyst, the reaction should be carried out in the lower temperature range; however, since conversion decreases with temperature, a lower space velocity or more active catalyst would be required to compensate for the loss in conversion. Also, a higher

Table I.   Effect of Process Variables[a]

Response	$1/S_V$	R	P	T
conv.	+s	+m	+m	+s
CO_2	+s	−m	0	+s
H_2O/CO_2	−s	+s	+m	−s
$(H_2/CO)_{RX}$	−s	+s	+m	−s
O/P	−m	−m	0	−s
H.C. sel.				
C_1	0	+s	0	+m
C_2–C_4	0	−m	0	−m
C_5^+	0	−m	0	+m
ROH	0	0	0	−m

[a] (+) increase, (−) decrease, (0) no effect. (m) mild relative effect over range studied, (s) strong relative effect over range studied.

H_2/CO ratio favors C_2–C_4, but unfortunately this has an adverse effect on hydrogen consumption and O/P ratio. It is obvious that an optimum condition represents a tradeoff of the various factors.

Kinetics. The data were fitted to a power rate law of the form:

$$\text{rate} = k\, P_H{}^m\, P_{CO}{}^n \tag{1}$$

where k is the global rate constant, P_H and P_{CO} are the partial pressures of H_2 and CO, and m and n are constants. Since the data did not show sharp curvature in the CO conversion vs. $1/S_V$ plots (*see* Figure 3), reasonably accurate initial slopes could be measured. On this basis, initial slopes could be related to total pressure and mole fractions of the inlet stream according to,

$$\text{slope} = \frac{x}{(1/S_V)} = k\, P^{m+n}\, (X_H{}^\circ)^m\, (X_{CO}{}^\circ)^{n-1} \tag{2}$$

where x is CO conversion, P is total pressure, and $X_H{}^\circ$ and $X_{CO}{}^\circ$ are initial H_2 and CO mole fractions. This analysis gave values of m and n of close to 1 and $-1/2$, respectively. Therefore, Equation 1 takes the form,

$$\text{rate} = k\, P_H/\sqrt{P_{CO}}. \tag{3}$$

To check the reasonableness of the values derived from initial slope data, the rate equation was formulated in terms of space time. Since hydrogen was not in great excess over CO in these runs, it was necessary to account for changes in hydrogen concentration as well as CO concentration over the reactor volume. For a fixed-bed reactor, Equation 3 now becomes,

$$k\sqrt{\frac{P}{X_{CO}{}^\circ}} \cdot \frac{1}{S_V} = \int_0^x \frac{\sqrt{1-x}}{R-ax}\, dx = Y \tag{4}$$

where R is the inlet H_2/CO ratio, a is the H_2/CO reacted, and Y represents the integral shown. Evaluation of the integral depends upon the ratio of R/a in a particular experiment, taking on the three forms:

$$Y(R > a) = \frac{2}{a}\left(1 - z + \sqrt{b}\left[\tan^{-1}\left(-\frac{1}{\sqrt{b}}\right) - \tan^{-1}\left(-\frac{z}{\sqrt{b}}\right)\right]\right) \tag{5a}$$

$$Y(R = a) = \frac{2}{a}(1 - z) \tag{5b}$$

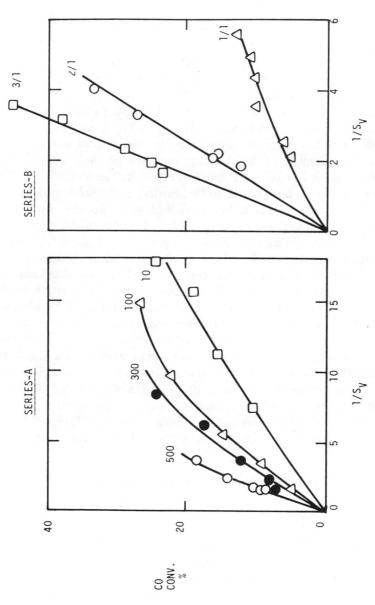

Figure 3. Effect of pressure and H₂/CO ratio on conversion. Series A, numbers are pressures in psig. Series B, numbers are H₂/CO ratios.

$$Y(R < a) = \frac{2}{a}\left(1 - z - 1/2\sqrt{-b}\left[\ln\frac{1 + \sqrt{-b}}{1 - \sqrt{-b}} - \ln\frac{z + \sqrt{-b}}{z - \sqrt{-b}}\right]\right)$$

(5c)

where $z = \sqrt{1 - x}$ and $b = (R - a)/a$. Since a and R were constant in each run, the appropriate expression of Y was used, and Y for each data point was plotted vs. $1/S_V$ to extract a value of k.

Values of k obtained by both the initial slope data and the integrated data are presented in Table II. Although some variation in k values were obtained, no consistent trends were observed, and the overall results are satisfactory within the experimental error. The rate constants show the catalyst in series B to be more active than in Series A, as was observed in the standard conversion data of Figure 2. We can only surmise that the pretreatment may have been more effective for the series B catalyst, as catalyst activity has been shown to be sensitive to the pretreatment (1).

A metal area for freshly reduced catalyst was estimated by oxygen chemisorption as described previously (1). A value of 6.3 mg/g was measured which, assuming one atom of oxygen adsorbed on each metal site, corresponds to 2.4×10^{20} adsorptive sites/g catalyst. (These values, as well as the k values from the kinetics are based on oven-dried catalyst weights; a weight loss of about 25% occurred in transformation to the reduced state at 500°C.) Assuming an average k of 3×10^{-3} cm³/g sec atm$^{1/2}$ at 250°C, an approximate turnover number of 1.8×10^{-3} sec^{-1} is obtained for reaction at 500 psig and H_2/CO of 2/1. Vannice (4) reports a turnover number of 2.8×10^{-2} sec^{-1} for methanation over a Co/Al_2O_3 catalyst at 275°C, 3/1, and 1 atm. Our value under comparable conditions (assuming an activation energy of 24 kcal/mol) is 4.4×10^{-3} sec^{-1}, or six times slower. Possible reasons for the discrepancy are: (1) the oxygen adsorption uptake may be high because of some bulk oxidation; (2) the metal surface may be rich in copper which is not active in the

Table II. Summary of Rate Constants

T(°C)	P (psig)	R	k_o[a]	k[a]	P (psig)	R	k_o[a]	k_o[a]
250	500	7/3	2.3	2.1	500	3/1	3.2	3.4
250	300	7/3	1.6	1.2	500	2/1	3.8	3.8
250	100	7/3	1.7	1.7	500	1/1	4.1	3.8
250	10	7/3	2.8	2.4	—	—	—	—
250	500	1/1	2.5	2.4	—	—	—	—
235	—	—	—	—	500	2/1	1.0	1.0
260	500	7/3	4.9	—	500	2/1	7.5	7.1
270	500	7/3	8.1	—	—	—	—	—

[a] $k \times 10^3$ cm³/g sec atm$^{1/2}$.

reaction but does adsorb oxygen; and (3) substantial coke may be present which covers active sites. However, a similar calculation on a 7.3% Co/Al_2O_3 catalyst reported previously (1) gave a rough turnover number of 6×10^{-3} sec⁻¹, about five times slower than Vannice's results. Therefore, reason 2 above does not appear to be the primary cause of lower activity for our catalysts.

The temperature variation could not be determined with sufficient accuracy with the data of series A or B, although a positive response was noted. An additional run was made with another catalyst of similar composition over a wider temperature range. Conditions were chosen to keep conversions low so that conversion data could be used directly in the analysis for activation energy. Unfortunately, this catalyst exhibited lower activity than the one used in series A and B, although selectivity distribution patterns were similar with both catalysts. An activation energy of 24 ± 2 kcal/mol was obtained for this catalyst.

Kinetic expressions similar to that of Equation 3 and similar activation energies have been reported for methanation over a cobalt–alumina catalyst (4) and for Fischer–Tropsch reaction over a cobalt–thoria catalyst (5). This similarity, despite appreciably different product distributions in the three cases, argues for a common rate-controlling step in the mechanisms.

It was noted earlier that plots of CO conversion vs. space time (Figure 3) were remarkably linear up to rather high conversions, which allowed analysis of the data from initial slopes. We also can analyze the data in terms of products formed. This was done for CO_2 and hydrocarbon yields, separately. Separation of reaction into these two components gives an interesting result, as shown in Figure 4. Here, it can be seen that the hydrocarbon formation shows more curvature on the plot than the overall CO conversion. The curve bends downward as would be expected from the kinetics, i.e., hydrocarbon formation should decrease from linearity with $1/S_V$ (with conversion). But the CO_2 formation curve bends upward. This effect is indicative of a series reaction, in which, as conversion increases, the H_2O formed reacts with CO to form CO_2 (water–gas shift reaction). This result is in agreement with the contention in the literature that H_2O formation is the primary reaction and that CO_2 forms by a secondary reaction (6).

Selectivities. Hydrocarbon selectivity data were treated to test conformance to Schulz–Flory polymerization kinetics as outlined by Henrici-Olive and Olive (7) for Fischer–Tropsch catalysis. The pertinent correlation is:

$$\log \frac{m_p}{P} = \log \frac{(1 - \alpha)^2}{\alpha} + P \log \alpha \tag{6}$$

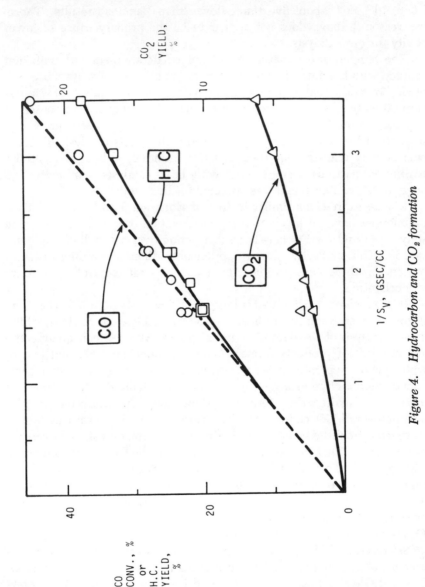

Figure 4.   Hydrocarbon and CO₂ formation

where m_p is the weight fraction of each carbon number fraction, P is the
carbon number, and α is the probability of chain growth. Equation 6 is
tested in Figure 5, where log m_p/P is plotted against P for one run
condition. The good fit signifies that this distribution is obeyed quite
well for our catalyst. Other run conditions gave equally good plots with
similar slopes. In all cases, the point for C$_2$ was low. Henrici-Olive and
Olive discuss reasons for this. The value of α is 0.55 from the slope and
0.52 from the intercept, indicating good conformance to Equation 6.
Further, we note that the C$_2$–C$_4$ fraction of 51% obtained in our run is
close to the predicted maximum achievable ·(56%) by this distribution
(7). We therefore conclude that our catalyst is close to maximum in
C$_2$–C$_4$ selectivity if this mechanism is operative.

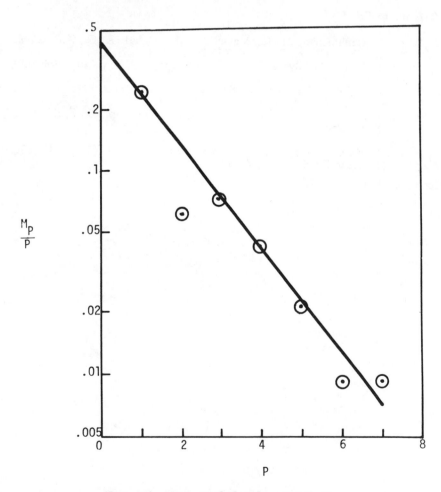

Figure 5.   Test of Schulz–Flory distribution

Promotional Effect of K₂O and Na₂O. Catalysts in which 0.05 and 0.1% K were added by impregnation to the oven-dried base catalyst gave identical CO conversion and selectivities as the unpromoted catalyst. It had been shown earlier that higher levels of sodium had a deleterious effect on conversion. These results indicate that addition of alkali is ineffective as a promoting agent for the $CoCu/Al_2O_3$ catalyst, in agreement with that reported for cobalt/thoria catalysts (8).

Acknowledgment

R. Jensen and B. Bailey assisted in parts of the work. This work was supported under ERDA contract E (49-18)–2006 and the State of Utah.

Literature Cited

1. Yang, C., ZamanKhan, M. K., Massoth, F. E., Oblad, A. G., *Am. Chem. Soc., Div. Fuel Chem., Prepr.* (1977) **22**(2), 148.
2. Anderson, R. B., "Catalysis," P. H. Emmett, Ed., Vol. 4, p. 109, Reinhold, New York, 1956.
3. Sinfelt, J. H., *Chem. Eng. Sci.* (1968) **23**, 1181.
4. Vannice, M. A., *Catal. Rev.—Sci. Eng.* (1976) **14**, 153.
5. Anderson, R. B., "Catalysis," P. H. Emmett, Ed., Vol. 4, p. 266 ff, Reinhold, New York, 1956.
6. Ibid, p. 280.
7. Henrici-Olive, G., Olive, S., *Angew. Chem., Int. Ed. Engl.* (1976) **15**, 136.
8. Pichler, H., *Adv. Catal.* (1952) **4**, 271.

RECEIVED August 14, 1978.

Cobalt-Based Catalysts for the Production of C₂–C₄ Hydrocarbons from Syn-Gas

A. L. DENT and M. LIN

Department of Chemical Engineering, Carnegie–Mellon University, Pittsburgh, PA 15213

Results of catalytic screening tests aimed at enhancing the selectivity of cobalt-based Fischer–Tropsch catalysts for the C_2–C_4 hydrocarbon fraction are reported. In these studies cobalt–zirconia, cobalt–copper, cobalt–chromia, and cobalt–manganese supported on kieselguhr or alumina were tested at 7.8 atm of a 1CO:3H$_2$:6He feed gas using a Berty, internally recycled, stirred catalytic reactor. The results to date show that while cobalt–chromia–kieselguhr and cobalt–zirconia–kieselguhr catalysts exhibit higher global rates for CO conversion (54.6 and 116.6 µmol/sec · gcat at 500 K, respectively), cobalt–manganese–alumina catalysts show the greatest selectivity potential to produce C_2–C_4 olefins in the 450–623 K range. While the optimum catalyst has yet to be established, the results demonstrate that alkali (as K_2O) plays a significant role in enhancing the selectivity.

Historically, the sources of petrochemical feedstocks have been related directly to the supply of petroleum. Initially, liquified petroleum gases (LPG) supplied this need. Later, as the supply of LPG diminished, naphthas became the primary source (*1*). The 1973 Arab oil embargo dramatically demonstrated the critical need for alternative sources of fossil carbon. Thus, any development to reduce our dependence on foreign suppliers of petroleum, e.g., an economic coal-based process, should have high priority (*2*). Hence, there has been a renewal of interest in Fischer–Tropsch processes (*3–14*), the catalytic synthesis of primarily C₅–C₁₁ hydrocarbons from CO–H₂ mixtures. Unfortunately, economic considerations indicate that a synthetic naphtha as cracker feedstock for

0-8412-0453-5/79/33-178-047$05.00/0

ethylene and propylene production does not appear to be favorable at today's market prices (*15, 16*). However, the direct synthesis of ethylene, propylene, and butene via Fischer–Tropsch synthesis does appear to be considerably more economically feasible. This requires a substantial improvement in the selectivity of the reaction leading to the C_2–C_4 olefins (*15, 18, 19*). This chapter reports some studies which are directed towards the achievement of this goal of more selective FT catalysts for the production of C_2–C_4 hydrocarbons from Syngas.

Experimental

Catalysts. Most of the catalysts used in these studies were prepared in our laboratory by the impregnation method (*6, 13*). Appropriate quantities of the metal nitrates were dissolved in sufficient amounts of distilled water to facilitate solution. To this solution the support material, alumina or kieselguhr, was added. The slurry was heated to 80°C for half an hour and was separated into the desired number of batches, e.g., alkali-free and one or two alkali-containing batches. To prepare the alkali batches, the desired quantity of potassium nitrate was added. The samples first were heated in an oven at 120°C to drive off excess water and then were calcined at 300°C for 12–16 hr in a furnace. Following calcination, the samples were crushed and screened to obtain material which was less than 250 μ (60 mesh) particle size. These samples were ready for installation in the reactor. Previous studies have indicated that catalysts prepared in this manner are equivalent to those prepared by coprecipitation techniques (*6*). Reagent grade $Co(NO_3)_2 \cdot 6H_2O$, $Cr(NO_3)_3 \cdot 9H_2O$, $Cu(NO_3)_2 \cdot 3H_2O$, $Th(NO_3)_4 \cdot 4H_2O$, KNO_3, $Fe(NO_3)_3 \cdot 9H_2O$, and a 50 wt % solution of $Mn(NO_3)_2$ were supplied by Fisher Scientific Company. Johns–Mansville's Celite Analytical Filter-Aid (kieselguhr), Conoco's SB Catapal Alumina (alkali content = 0.004% as Na_2O), and Alcoa's Activated Alumina F1 (alkali content = 0.9 wt %) were used as support materials. Activated carbon was used as support for one Co–Mn catalyst and was supplied by Calgon Corporation. Table I summarizes the catalysts used in these studies.

Catalytic Reaction Unit. Figure 1 represents a catalytic reactor unit designed for Fischer–Tropsch and methanation catalyst screening tests. The unit is capable of operating in the temperature range 100°–450°C and a pressure range 0–200 psig. The CRU consists of a gas delivery system, an internally recycled catalytic reactor, liquids-separation traps, and an analytical system.

The individual gases are metered by Nupro "very fine" metering values (Crawford Company) and are measured using a wet-test meter at NTP. A six-port valve downstream of the gas blending point permits selection of feed gas or purge gas to be sent to the reactor. (During flow calibration, helium is sent to the reactor.) Downstream of the reactor is a series of traps at varying temperature which are used to collect: (a) heavy oils (75°C), (b) water and alcohols (0°C), and (c) light oils (− 70°C) so that only C_5 and lower hydrocarbons are sent to the analytical system. The unit is maintained at the desired pressure (usually 100 psig) by a Fairchild Model 10BP back pressure regulator.

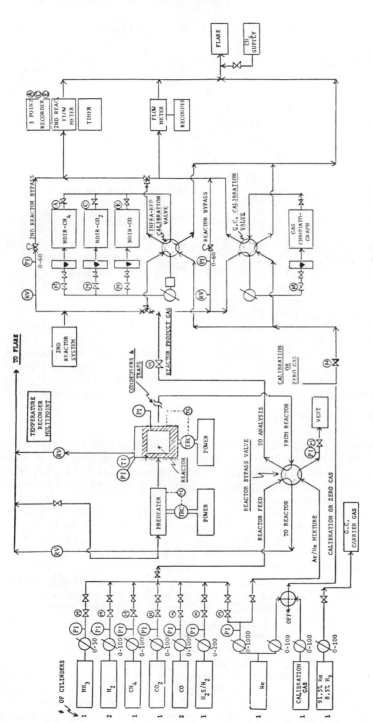

Figure 1. Diagram of catalytic reactor unit and analytical system

Table I. Summary of F–T Catalysts

Item No.	Catalyst Composition	Comments
1	$100Co:5.0ZrO_2:200Kgh$[a]	Commercial catalyst obtained from Chemetron Corp. (Girdler G67)
2	$100Co:10Cu:18ThO_2:200Kgh$	Alkali free, coprecipitated with $(NH_4)_2CO_3$
3	$100Co:10Cr:200Kgh$	
4	$100Co:10Mn:200Kgh$	
5	$100Co:10Mn:200$ Act. Carbon	
6	$100Co:50Mn:400Al_2O_3$	Prepared from Conoco Catapal Al_2O_3
7	$100Co:50Mn:400Al_2O_3:5.5K_2O$	Prepared from Conoco Catapal Al_2O_3
8	$100Co:10Mn:400Al_2O_3:3.6K_2O$	Prepared from Alcoa Al_2O_3, F-1
9	$100Co:10Mn:400Al_2O_3:5.0K_2O$	Prepared from Alcoa Al_2O_3, F-1
10	$100Co:50Mn:600Al_2O_3:20.4K_2O$	Prepared from Alcoa Al_2O_3, F-1
11	$100Co:50Mn:600Al_2O_3:35.4K_2O$	Prepared from Alcoa Al_2O_3, F-1
12	$100Co:20Mn:500Al_2O_3:6.5K_2O$	
13	$100Co:100Mn:700Al_2O_3:18.0K_2O$	
14	$20Co:100Mn:500Al_2O_3:12.4K_2O$	
15	$100Co:20Mn:500ZnO:2K_2O$	
16	$100Fe:20Mn:400Al_2O_3:2K_2O$	Prepared from Alcoa Al_2O_3, F-1
17	$100Fe:100Mn:668Al_2O_3:2K_2O$	Prepared from Alcoa Al_2O_3, F-1
18	$100Fe:50Mn:500Al_2O_3:2K_2O$	Prepared from Alcoa Al_2O_3, F-1
19	$20Fe:100Mn:400Al_2O_3:2K_2O$	Prepared from Alcoa Al_2O_3, F-1

[a] Kgh = Kieselguhr support.

The reactor used for these studies is the Autoclave Engineer's Berty internally-recycled reactor (19) which is supplied with a 3-kW heater. Temperature control is achieved via a Nanmac Model PC-1 temperature controller which operates a mercury relay switch–variac combination which powers the heaters.

Two modifications of the Berty reactor (19) permitted greater temperature control and more accurate temperature measurements. The first modification involved installation of a 1/8-in. diameter coil in the head of the reactor, through which pressurized air could flow. The air flow through the coil was controlled by a relay switch that was connected to a separate Nanmac temperature controller. The second modification was simply to replace the upper thermocouple by dual chromel–alumel thermocouples (supplied by Thero-Electric) which extended downward into the catalyst bed. One of these thermocouples recorded the bed temperatures as part of a 12-temperature profile, while the other was used to operate the air coil's controller.

Product Analysis. Analysis of gases from the CRU is achieved in two ways. Carbon monoxide, methane, and carbon dioxide are analyzed continuously by Mine Safety Appliances' Model LIRA 303 nondispersive analyzers (NDIR), while the total effluent gas analysis is obtained by gas chromatography using a combination of porapak QS and charcoal columns operated isothermically at 100°C and 60 cm³/min helium-carrier-flow rate in two Perkin–Elmer Model 154C instruments containing TC detectors. Sample injection for the GC analysis (5 cm³ loops) occurs via automated Valco gas-sampling valves which are controlled by a Varian CDS101 Data Analyzer. The analysis, which is completed in 20 min, is quantitatively detailed only for the C_1 to C_4 hydrocarbon fraction in this study. Liquid products should accumulate in the three traps which can be inspected after each run.

Gases. Carbon monoxide (99.8%) and hydrogen (99.95%) gases used in these studies were supplied by AIRCO; helium was obtained from North American Cryogenic Company. A calibration gas for the NDIR and gas chromatograph containing CO, CO_2, CH_4, C_2H_4, C_2H_6, C_3H_8, and He was supplied by Matheson. A second calibration gas containing C_4H_8, C_4H_{10}, and He balance also was used.

Screening Tests. For most studies, 35 cm³ of catalyst were charged to the reactor with adequate quantities of borosilicate-glass wool to position vertically the catalyst bed in the center of the reactor. In situ reduction was achieved using 500 cm³/min hydrogen at 400°C and 35 psig for at least 16 hr. (A mixture of hydrogen and helium was used to bring the catalyst bed from ambient temperatures to the reduction temperature.) Following reduction, the hydrogen and helium flow rates were adjusted to the desired values. Using the CO-NDIR, the CO flow rate was adjusted stepwise over a six-hour period to its final value. At this point, the feed gas composition was $1CO:3H_2:6He$. This conditioning period was found to assure reproducible activity for all of the catalysts studied (*20*).

Following preconditioning, the product gases were measured at several temperatures in a manner illustrated in Figure 2 with each temperature being maintained for at least 4 hr. This procedure permitted an assessment of catalyst aging characteristics, reproducibility, and per-

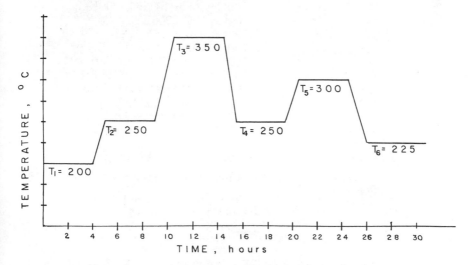

Figure 2. Temperature excursion during screening tests

formance at various temperatures. At the beginning of each new temperature setting, the RPM was varied to determine the conditions at which mass-transfer interferences became important. An RPM setting (internal recycle ratio) was used which was at least 30% greater than the point at which the conversion became independent of RPM. Thus, an activation energy for CO conversion could be obtained which should be free of mass-transport interferences.

Results and Discussion

The results of the screening tests of cobalt catalysts investigated in this study are summarized in Tables II, III, IV, and V and Figure 3. From this summary, several features are apparent and warrant comment.

Activity. A comparison of the global rates of CO conversion on a per gram of catalyst or on a per gram of cobalt in the catalyst at 500 K shows that the activities of the chromium- and zirconium-doped catalysts were substantially higher than any of the other catalysts studied. (Specific rates on a per active catalyst site basis (13, 21) are not available for these catalysts. Such measurements will be undertaken for the more promising catalysts in the near future (22). Justification for this use of the continuous stirred-tank reactor (CSTR) design equation was provided by pulse tracer experiments (20).) These are followed by the activated carbon-

Table II.   Carbon Monoxide Conversion

Run No.	Catalyst
19	$100Co:10Cu:18ThO_2:200Kgh$
28	$100Co:10Cr:200Kgh$
29	$100Co:5.0ZrO_2:200Kgh$[a]
30	$100Co:10Mn:200Kgh$
31	$100Co:10Mn:200$ Act. Carbon
24	$100Co:10Mn:400Al_2O_3:3.6K_2O$
32	$100Co:10Mn:400Al_2O_3:5.0K_2O$
34	$100Co:50Mn:400Al_2O_3$
35	$100Co:50Mn:400Al_2O_3$
Vannice	$10Co:450Al_2O_3$[f]
Bartholomew	$10Co:400Al_2O_3$[g]
Massoth	$100Co:137.5Cu:1325Al_2O_3$[h]

[a] All studies were conducted at 7.80 atm (100 psig) using a 10% CO, 30% H_2, 60% He feed unless otherwise noted.
[b] Calculated from experimental rate data assuming Arrhenius law behavior.
[c] Rate is defined at temperature $T(K)$ for CSTR behavior as

$$\text{rate} = \frac{X_{CO_T} \cdot Q_t \cdot C_{CO_f}}{W_{cat}}$$

where Q_t is volumetric flow rate of feed, X_{CO_T} is conversion at $T(K)$, C_{CO_f} is the molar concentration of CO in feed, and W_{cat} is weight of catalyst.

supported Co–Mn catalyst. For the alumina supported Co–Mn catalysts, the addition of alkali has no drastic effects on the global rates over the ranges studied in this work. The global rates for the Co–Mn · Al_2O_3 catalysts appear to be comparable with the "Co–Al_2O_3–100" catalyst of Bartholomew (23) and the Co–Cu–ThO_2 catalyst of Massoth (18).

As indicated by Figure 2, screening tests were conducted in a manner which permitted an assessment of the stability of the catalyst as a function of time on stream. Generally, under the conditions described earlier, the activity at T_4 was within 10% of that at T_2 where T_2 and T_4 represent equivalent ($\pm 2°C$) test-run temperatures which occurred early and late in the run, respectively. Hence, severe catalyst deactivation usually was not observed.

Table II also lists the apparent activation energies for CO conversion obtained for these catalysts. Our values are comparable with those reported by others in recent literature (13, 14, 23, 24). The value of E_{act} for Co–Mn on kieselguhr seems unusually low by comparison with other values. This anomaly will be pursued in future work using the Koros–Kowak technique (26), as a means of determining whether this low value is attributable to inter-particle mass-transport effects or whether a compensation effect might be operative (13, 14).

Activity of Cobalt F–T Catalysts at 500 K[a]

Global Rate $\times 10^{6}$[b,c] (Mol CO converted/ sec · gcat)	Global Rate $\times 10^{6}$[c,d] (Mol CO converted/ sec · g · Co)	Apparent Activation Energy, E_{CO} (kcal/mol)
2.59	8.5	33.2 ± 3.0
54.6	169.3	36.3 ± 2.0
116.6	356.0	32.0 ± 1.5
5.72	17.7	11.6 ± 0.6
19.4	60.1	14.7 ± 3.1
3.06	15.7	30.5 ± 2.0
4.88	25.1	30.0 ± 2.7
1.44	7.9	21.4 ± 2.0
1.29	7.2	26.6 ± 2.2
—	—	26.7 ± 6.2
1.19	6.0	28.0
~ 18.1	283.0	—

[d] Gram cobalt based on impregnated catalyst formulation.
[e] Commercial Girdler catalyst diluted 1:1 by volume with kieselguhr.
[f] This result is from the work of M. A. Vannice (13). The alkali content of the alumina support was not specified by the author. Pressure was 1 atm in a microcatalytic reactor.
[g] This result is from the work of C. H. Bartholomew (23). J. Kaiser SAS (5 × 8 mesh) alumina pellets were impregnated with aqueous cobaltous nitrate. Pressure was 24.8 atm in a fixed-bed reactor using 1% CO, 4% H_2, 95% N_2 feed.
[h] This result is from the work of Oblad and Massoth (16). Na_2CO_3 precipitated catalysts were used at 52 atm in a fixed-bed reactor with $3H_2/1CO$ feed gas.

Run # 19

100 Co: 10 Cu: 18 ThO_2: 200 Kghr

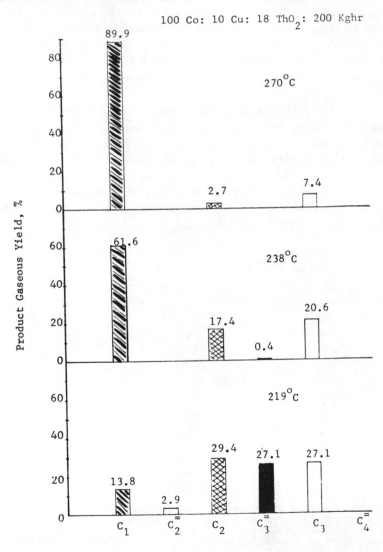

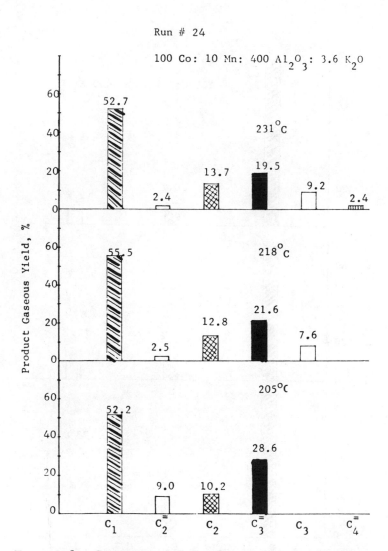

Figure 3a,b.   Comparison of F–T gaseous product distributions for cobalt catalysts, carbon-atom % ($H_2/CO = 3.0$ for all runs, 7.8 atm (100 psig). (a) (left) Run No. 19. (b) (above) Run No. 24.

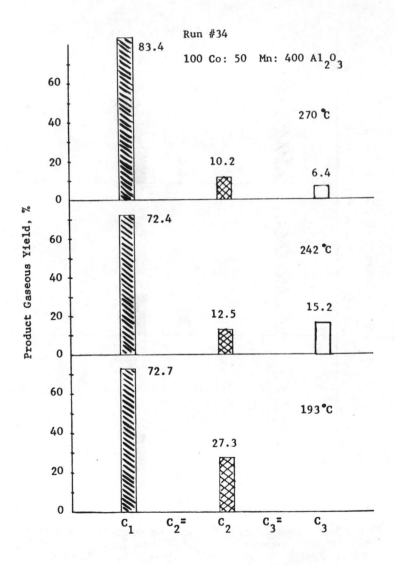

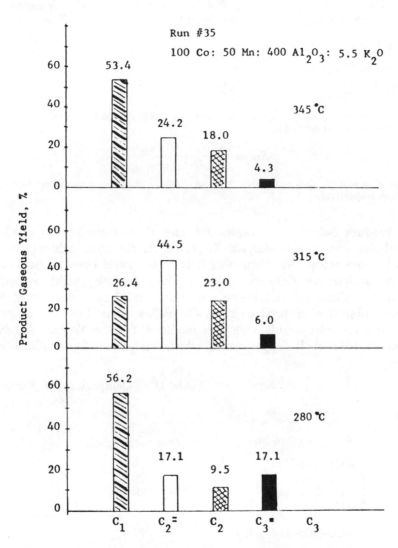

Figure 3c,d.   Comparison of F–T gaseous product distributions for cobalt catalysts, carbon-atom % ($H_2/CO = 3.0$ for all runs, 7.8 atm (100 psig). (c) (left) Run No. 34. (d) (above) Run No. 35.

Table III. Comparison of Product Distributions

Run No./Catalyst	Sp. Vel (cm³/sec · gcat)	CO % Conv.[b] (−)	Reactor Temp (K)
19/Co–Co–Kgh	1.06	{ 5.9 { 23.6	489 509
28/Co–Cr–Kgh	36.0	34.2	520
29/Co–ZrO₂–Kgh	50.3	26.7	512
30/Co–Mn–Kgh	26.5	3.2	548

[a] All studies were conducted at 7.80 atm using a 10% CO, 30% H₂, 60% He feed.
[b] Conversion is defined as

$$X_{co} = \frac{Q_f Y_{co_f} - Q_p Y_{co_p}}{Q_f Y_{co_f}}$$

where Y is mol %, Q is volumetric flow rate, subscripts f and p represent feed and product, respectively.

Product Selectivity. Tables III and IV summarize the product distributions for these catalysts. In Table III the kieselguhr-supported catalysts are compared. From this it is to be noted that the two most active catalysts for CO conversion are, unfortunately, highly selective towards methanation and hence are poor candidates for meeting the desired objective of producing C_2–C_4 olefins. The Co–Mn catalysts, however, are comparable with the modified Fischer–Tropsch catalyst (Run No. 19), cobalt–copper–thoria, which was prepared as an alkali-free

Table IV. Comparison of Product

Run No./Catalyst	Sp. Vel. (cm³/sec · gcat)	CO % Conv.[b] (−)
30/Co–Mn–Kgh	26.5	3.2
31/Co–Mn–C	20.4	19.7
34/Co–Mn–Al₂O₃	0.42	36.0
24/Co–Mn–Al₂O₃–K₂O	1.27	12.4
32/Co–Mn–Al₂O₃–K₂O	27.4	{ 24.4 { 4.5 { 2.0
35/Co–Mn–Al₂O₃–K₂O	0.42	{ 16.6 { 8.2 { 3.0
49/Co–Mn–Al₂O₃–K₂O[c]	0.82 0.44	{ 5.6 { 12.2
44/Co–Mn–ZnO–K₂O	0.73	63.4

[a] All studies were conducted at 7.80 atm using a 10% CO, 30% H₂, 60% He feed.
[b] See Table III for definition of terms.

for Kieselguhr-Supported Cobalt F–T Catalysts[a]

	Productive Selectivity Carbon Atom %			
% CO_2	% CH_4	% C_2-C_4	C_{5+}[d]	O/P[e]
1.1	4.8	30.1	64.0	0.5
21.9	27.8	28.0	22.4	0.3
26.7	72.2	0.2	1.0	0.2
30.8	67.8	0.5	0.9	0
7.8	55.4	10.6	26.2	1.3

[e] Olefin to paraffin ratio is based on carbon atom % C_2–C_4 olefins per carbon atom % C_2–C_4 paraffins.

[d] C_5^+ represents carbon balance but may contain alcohols as well.

reference catalyst. Moreover, at comparable conversion levels, the olefin/paraffin ratio is higher for the Co–Mn entry.

In Table IV, the effects of support and alkali on product selectivity are illustrated for the Co–Mn series. The role of the support is not obvious in these results. However, it is seen that for those entries which contain alkali (K_2O), the selectivities for C_2–C_4 olefins are higher. Figure 3 provides a visual comparison of the product distribution for the 100Co:50Mn–400Al$_2$O$_3$:5.5K$_2$O catalyst (Run No. 35) to its alkali-free analog

Distributions for Co–Mn F–T Catalysts[a,b]

Reactor Temp. (K)	Product Selectivity Carbon Atom %				
	% CO_2	% CH_4	% C_2-C_4	C_{5+}	O/P
548	7.8	55.4	10.6	26.2	1.3
548	35.5	59.9	4.6	0	0
515	6.8	49.7	18.9	24.6	0
491	3.4	26.5	12.1	58.0	1.4
546	27.8	46.9	25.3	0	1.2
526	29.5	65.5	5.0	0	3.4
512	32.3	33.1	8.7	25.8	> 10
618	35.5	33.8	30.8	0	1.6
585	28.1	19.9	56.7	0	2.2
552	20.5	9.8	8.0	61.7	4.2
552	48.4	14.7	25.0	11.9	6.2
552	32.7	15.3	47.9	4.1	3.8
602	34.5	44.1	17.2	4.2	0.4

[c] Results taken from Ref. *22.*

Table V. Comparison of Olefin

Catalyst	Sp. Vel (cm^3/sec · gcat)
$100Co-50Mn-400Al_2O_3-5.5K_2O$	0.42
$100Co-50Mn-600Al_2O_3-12.9K_2O$	0.79
$100Co-50Mn-600Al_2O_3-20.4K_2O$	0.39
$100Co-50Mn-600Al_3O_2-35.4K_2O$	0.44
$100Co-10Cu-200Kgh-18ThO_2$	1.06
$100Co-100Kgh-18ThO_2-(K_2CO_3$ ppd)	~ 0.02
	~ 0.7
$100Co-16ThO_2-88Kgh$ (Na_2CO_3 ppd)	0.60
$100Co-137.5Cu-1325Al_2O_3$ (Na_2CO_3 ppd)	0.39
$100Fe-100Mn-668Al_2O_3-7.3K_2O$	1.15
Zein El Deen et al. (Catalyst A–Fe, Mn, ZnO + K_2O)	~ 0.2
Sasol entrained iron catalyst	—
Oblad et al. (5Mn/100Fe)	1.06
Amoco ppd iron catalyst	~ 0.2

[a] Olefin efficiencies are based on CO_2- and H_2O-free yields and total conversion to $C_2^=$–$C_4^=$, respectively.
[b] Studies were conducted at 7.80 atm using a 10% CO, 30% H_2, 60% He feed in a Berty reactor.
[c] A fixed-bed reactor operating at 1 atm with a 33.3% CO and 66.7% H_2 feed was used.
[d] Studies were conducted at 9.16 atm using a 30% CO and 70% H_2 feed in a fixed-bed reactor.
[e] A fixed-bed reactor operating at 52 atm of a 25% CO and 75% H_2 feed gas was used.

(Run No. 34) and to those from Run No. 19 and No. 24. Table V provides a comparison of the C_2–C_4 olefin efficiencies for the Co–Mn catalysts from this laboratory (23) to those efficiencies calculated from product distributions found in the literature (7, 9, 10, 17, 18, 24, 25). Where the data were available, olefin efficiencies were calculated on the basis of hydrocarbon selectivity (on a CO_2-free basis, column No. 4) and also on a productivity per feed carbon basis (column No. 5). The latter basis takes into account changes in the olefin/paraffin ratio and catalyst activity for variation of temperature and space velocity. In addition to demonstrating the very significant effects of alkali (K_2O), such comparisons seem to support the enthusiasm of the authors for the Co–Mn–Al_2O_3–K_2O catalyst systems for the production of light olefins from synthesis gas. Work is in progress to determine the optimum composition and alkali content of Co–Mn catalysts as well as a similar Fe–Mn series of alumina-supported catalysts.

The enhanced selectivity of Fischer–Tropsch catalysts with increasing alkali content has been the subject of a number of reports in the literature

Efficiencies for Fischer–Tropsch Catalysts

Reactor Temp. (K)	*C_2-C_4 Olefin Efficiency*[a] *$C_2^=$–$C_4^=$ C Atoms 100 HC Atoms in Products*	*$C_2^=$–$C_4^=$ C Atoms 100 C Atoms in Feed*	*Ref.*
585	54.2	4.45	This work[b]
518	22.8	1.73	*22*[b]
538	41.2	4.51	*22*[b]
552	56.3	4.62	*22*[b]
509	7.6	1.40	This work[b]
463	8.1	$\leq$ 5.1	*9,10*[c]
463	9.7	$\leq$ 1.0	
498	1.58	1.24	*17*[d]
508	9.8	2.22	*18*[e]
636	15.2	2.64	*22*[f]
529	27.0	4.51	*24*[g]
593	25.3	—	*10*[h]
473 (?)	~ 37.0	~ 2.6	*25*[i]
588	27.0	—	*7*[j]

[f] Studies were conducted at 7.80 atm using a 15% CO, 45% H_2, 40% He feed in a Berty reactor.

[g] Reactor conditions were 10 atm using a feed gas containing 40.1% CO, 39.3% H_2, and 240.4% Ar. Data represent products after 8 hr on stream in a Berty reactor.

[h] An entrained-bed recator operating at 22 atm of 33.3% CO and 66.7% H_2 feed was used.

[i] Conditions were 52 atm of 33.3% CO and 66.7% H_2 feed gas in a fixed-bed reactor.

[j] A fluidized-bed reactor with a recycle ratio of 2 and 33.3% CO and 66.7% H_2 feed gas was pressurized to 18 atm for this study. Total CO conversion was not reported.

(*15, 16, 25, 27*). Recently, Yang and Oblad (*25*) reported that the olefin/paraffin ratio in the C_2–C_4 formation increased to higher levels at around 0.2 g K/100 g Fe, then leveled off as the potassium content increased further. Although methane decreased correspondingly as the potassium content increased from zero to 0.2 g K/100 g Fe, the C_2–C_4 fraction also decreased slightly. Between 0.2 and 0.5 g K/100 g Fe, they observed little effect of potassium on the C_2–C_4 fraction or on the C_{5+} fraction. Earlier, Dry and Oosthuizen (*27*) had studied the effect of surface basicity on hydrocarbon selectivity in the Fischer–Tropsch synthesis over a number of magnetite catalysts. They concluded that: (a) the surface basicity correlated well with hydrocarbon selectivity, the higher the basicity the higher the long-chain hydrocarbon selectivity; and (b) the surface basicity and hence hydrocarbon selectivity depended not only on the amount of K_2O present but also on how well it was distributed over the catalyst surface. Assuming that this is also the case for other metal systems, we should strive to obtain the highest possible metal and alkali dispersions possible. This is more likely to be achieved by use of high surface area supports such as alumina rather than low area kieselguhrs. Since Oblad's (25) catalysts were prepared by coprecipitation and are probably low area (28), this may account for the lack of sensitivity to potassium above 0.2 g K/100 g Fe, as observed by Anderson (29). That is, their surfaces may have become saturated by potassium; hence, no further effects could be observed. We have observed substantially improved selectivity to C_2–C_4 olefins for Co–Mn catalysts with addition of potassium as shown in Tables IV and V. However, we recognize that the optimum system has yet to be achieved.

Recently, Büssemeir, Frohning, and Cornils (15) presented their concepts for design of new FT catalysts which have increased selectivity for the lower olefins, C_2–C_4 (30, 31, 32). They argued that the "ideal catalyst" should show a moderate affinity for both carbon monoxide and hydrogen adsorption. Based on the bonding energy data for both hydrogen and carbon monoxide on the first row of transition metals (33), they concluded that manganese should be superior to iron which is in turn superior to nickel and cobalt. Hence, combinations of vanadium or titanium with iron and/or manganese should provide "promising catalysts (30, 31, 32, 33). They found, however, that while iron-free catalysts did not prove successful for FT synthesis, doping of iron catalysts with varying concentrations of metals from sub-groups IV up to VII resulted in a considerable limitation of the product distribution and a relatively high yield of short-chain olefins (15). Similar results have been reported by Zein El Deen, Jacobs, and Baerns (24). For comparison, their "Catalyst A" (which is probably an Fe–Mn based catalyst) is included in Table V. Their results using $H_2/CO = 1$ compare very favorably with our results which represent much higher H_2/CO ratios where the amount of olefins is expected to be less (22). While detailed comparisons between their study and our work are difficult to make, the results to date with cobalt catalysts are sufficiently encouraging to warrant further efforts to obtain catalysts with improved olefin selectivity. Studies which include Co–Mn, Fe–Mn, and Fe–Mn–Co combinations supported on alumina and doped with potassium oxide (see Table I, items 11–19) are in progress.

Acknowledgment

The authors wish to acknowledge the Department of Energy for support of the work under ERDA Contract No. E(49-18)-1814.

Literature Cited

1. Lambrix, J. R., Morris, C. S., *Chem. Eng. Prog.* (1972) **68**(8), 24.
2. "Coal Gasification Technology at Center Stage," *Chem. Eng. News* (Jan. 10, 1972) 36–38.

3. Fischer, F., Tropsch, H., *Brennst.-Chem.* (1923) **4**, 276.
4. Ibid. (1924) **5**, 201, 217.
5. Ibid. (1926) **7**, 97.
6. Storch, H. H., Golumbic, N., Anderson, R. B., *"The Fischer–Tropsch and Related Syntheses,"* Wiley, New York, 1951.
7. Weitkamp, A. W., Seelig, H. S., Bowman, N. J., Cady, W. E., *Ind. Eng. Chem.* (1953) **45**(2), 343.
8. Neale-May, W. M., *S. Afr. Ind. Chem.* (May 1958) 80–93.
9. Pichler, H., Schulz, H., Elstner, M., *Brennst.-Chem.* (1967) **48**, 78.
10. Pichler, H., Schulz, H., *Chem. Ing. Tech.* (1970) **42**, 1162.
11. Dry, M. E., Shingles, T., Boshoff, L. J., *J. Catal.* (1972) **25**, 99.
12. Mills, G. A., Steffgen, F. W., *Catal. Rev.* (1973) **8**, 189.
13. Vannice, M. A., *J. Catal.* (1975) **37**, 449, 462.
14. Ollis, D. F., Vannice, M. A., *J. Catalysis* (1975) **38**, 515.
15. Büssemeier, B., Frohning, C. D., Cornils, B., *Hydrocarbon Process.* (Nov. 1976) 105–112.
16. Shah, Y. T., Perrotta, A. J., *Ind. Eng. Chem., Prod. Res. Dev.* (1976) **15**(2), 123.
17. Zaman Khan, M. K., Yang, C. H., Oblad, A. G., *Am. Chem. Soc., Div. Fuel Chem., Prepr.* (1977) **22**(2), 138.
18. Yang, C. H., Zaman Khan, M. K., Massoth, F. E., Oblad, A. G., *Am. Chem. Soc., Div. Fuel Chem., Prepr.* (1977) **22**(2), 148.
19. Berty, J. M., *Chem. Eng. Prog.* (1974) **70**, 78.
20. Lin, M., M.S. Thesis, Carnegie-Mellon University (1976).
21. Vannice, M. A., *Catal. Rev.—Sci Eng.* (1976) **14**, 153.
22. Dent, A. L., Halemane, K. P., *"Cobalt–Manganese Catalysts for Selective Fischer–Tropsch Synthesis,"* N. Am. Meeting of Catal. Soc., 6th, Chicago, March, 1979, unpublished data.
23. Bartholomew, C. H., *"Alloy Catalysts with Monolith Supports for Methanation of Coal-Derived Gases,"* Final Report to U.S. DOE, Contract #EX-76-S-01-1790 (September 1977).
24. Zein El Deen, A., Jacobs, J., Baerns, M., *ACS Symp. Ser.* (1978) **65**, 26–36.
25. Yang, C. H., Oblad, A. G., *Am. Chem. Soc., Div. Pet. Chem., Prepr.* (1978) **23**(2), 513–520.
26. Koros, R. M., Nowak, E. J., *Chem. Eng. Sci.* (1967) **22**, 470.
27. Dry, M. E., Oosthuizen, G. J., *J. Catal.* (1968) **11**, 18.
28. Phung, L., M.S. Thesis, Carnegie-Mellon University (1978).
29. Anderson, R. B., *Catalysis (1954–1960)* (1956) **4**, Ch. 1–4.
30. German Patent Applications DOS 2.507.647 (Feb. 19, 1975).
31. German Patent Applications DOS 2.518.964 (April 29, 1975).
32. German Patent Applications DOS 2.536.488 (August 16, 1975).
33. Kolbel, H., Tillmetz, K. O., *Ber. Bunsenges. Phys. Chem.* (1972) **11**, 1156.

RECEIVED June 22, 1978.

Hydrogenation of CO and CO_2 on Clean Rhodium and Iron Foils

Correlations of Reactivities and Surface Compositions

D. J. DWYER[1], K. YOSHIDA[2], and G. A. SOMORJAI[3]

Materials and Molecular Research Division, Lawrence Berkeley Laboratory, Department of Chemistry, University of California, Berkeley, CA 94720

The apparatus that permits UHV surface characterization and high-pressure (1–20 atm) catalytic reactions to be carried out was used to investigate the hydrogenation of CO over iron and rhodium surfaces. Small-surface-area (~ 1 cm^2) metal samples were used to catalyze the H_2/CO reaction at high pressures (1–6 atm). Surface compositions of the metal samples were determined before and after the reaction, and the results were correlated with the observed product distributions and reaction rates. Differences in poisoning characteristics and product distributions indicate the importance of additives in controlling activity and selectivity. The addition of olefins to the feed have markedly changed the product distribution over the iron catalyst, indicating the major role that readsorption and secondary-surface reactions play in controlling the product distribution.

Studies of catalyzed reactions of CO and CO_2 with hydrogen to produce hydrocarbons have had a profound effect on the chemical research and chemical technology (1, 2, 3). As a result of coal gasification (coal + H_2O → CO + H_2), CO and H_2 are produced and may be used as feedstock or as a fuel (water gas). Through the use of the water shift reaction (CO + H_2O ⇌ CO_2 + H_2), the CO–H_2 mixture can be en-

[1] Current address: Exxon Research and Engineering Labs., Linden, NJ 07036.
[2] Current address: Catalysis Institute, Hokaido University, Japan.
[3] Author to whom correspondence should be sent.

Table I.

Water Gas Reaction	Shift Reaction
$C(graphite) + H_2O \leftrightarrows CO + H_2$	$CO + H_2 \leftrightarrows CO_2 + H_2$
$\Delta H_{500\ K} = +32.0\ kcal\ mol^{-1}$	$\Delta H_{500\ K} = -9.5\ kcal\ mol^{-1}$
$\Delta G_{500\ K} = +15.2\ kcal\ mol^{-1}$	$\Delta G_{500\ K} = -4.8\ kcal\ mol^{-1}$
$\Delta G_{1000\ K} = -1.9\ kcal\ mol^{-1}$	$\Delta G_{1000\ K} = -0.6\ kcal\ mol^{-1}$

riched with hydrogen that is desirable in many of the chemical reactions of these two molecules. Table I lists the thermodynamic data for the coal gasification and water shift reactions.

With various ratios of CO and H_2, the production of hydrocarbons of different types are all thermodynamically feasible. Let us consider the formation of alkanes, alkenes, and alcohols according to Reactions 1, 2, and 3. The standard free energies of formation of the various products

$$(n + 1)\ H_2 + 2n\ CO \rightarrow C_nH_{2n+2} + n\ CO_2 \tag{1}$$

$$2n\ H_2 + n\ CO \rightarrow C_nH_{2n} + n\ H_2O \tag{2}$$

$$2n\ H_2 + n\ CO \rightarrow C_nH_{2n+1}OH + (n - 1)\ H_2O \tag{3}$$

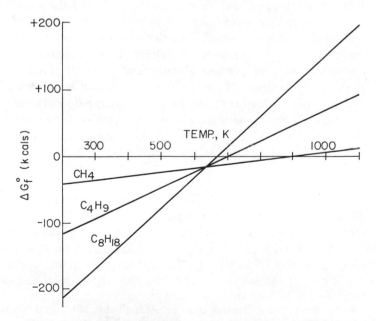

Figure 1. Temperature dependence of the free energy of formation of alkanes from H_2 and CO

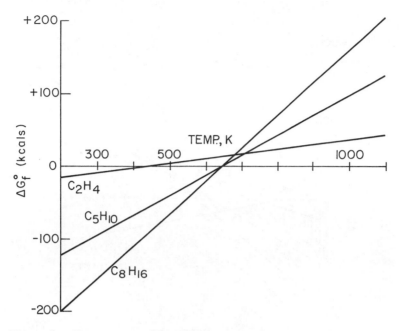

Figure 2.   Temperature dependence of the free energy of formation of alkenes from H$_2$ and CO

as a function of temperature are shown in Figures 1, 2, and 3. Since these are exothermic reactions, low temperatures favor the formation of the products. However, the reactions are all kinetically limited. (They have low turnover numbers, 10^{-2} to 10 molecules/surface atom sec.) Therefore, higher temperatures in the range 500–700 K usually are used to optimize the rates of formation of the products. According to the LeChatelier principle, high pressures favor the association reaction that is accompanied by a decrease in the number of moles in the reaction mixture as the product molecules are formed. Thus the formation of higher-molecular-weight products is more favorable at high pressures. Figures 4 and 5 show that pressures in excess of 20 atmospheres are desirable to produce higher-molecular-weight alcohols or benzene. If reactions are carried out at one atmosphere, for example, the catalyst cannot exhibit its real performance because of thermodynamic limitations. Thus, we must use a high pressure batch or flow reactor capable of carrying out the reactions of CO/H$_2$ mixtures up to 100 atmospheres.

Chemical reactions that produce methane from CO and H$_2$ are called methanation reactions. The other reaction that produces a C$_1$ hydrocarbon yields CH$_3$OH, methanol. All other reactions that produce C$_2$-C$_n$ hydrocarbons are called Fischer–Tropsch reactions, named after the scientists who developed much of the early CO/H$_2$ chemistry. In

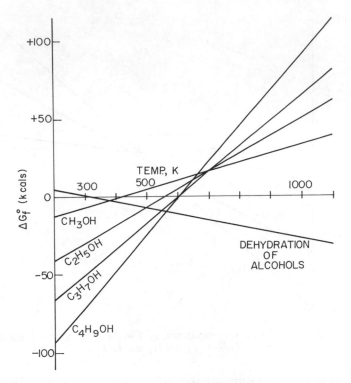

Figure 3. *Temperature dependence of the free energy of formation of alcohols from H₂ and CO*

Germany during the second World War, almost all of the gasoline and much of the hydrocarbon chemicals were produced by the Fischer–Tropsch reaction because of the lack of crude oil. The typical product distribution that was obtained by using a cobalt catalyst that was promoted with thorium and potassium oxides (ThO_2 and K_2O) is shown in Table II. Most of the products are straight-chain hydrocarbons and the C_4-C_9 fraction that is used in gasoline is of low octane number. It is unfortunate that this product distribution is compared with those obtained by the conventional hydrocracking and reforming processes from crude oil since it only reflects our level of understanding of catalytic chemistry of 40 years ago. There are several chemical processes that play important roles in the present chemical technology which use CO and H_2 mixtures to yield selectively the desired products. Methanol can be produced with excellent yield over zinc chromate, copper chromate catalysts that exhibit both the necessary hydrogenation and oxidation activities ($CO + 2H_2 \rightarrow CH_3OH$) (4). Recently, palladium and platinum also were found to carry out this reaction, selectively, at high pressures (12 atm) (5).

Methane is produced with a relatively high rate, selectively, over nickel catalysts. This process also finds industrial applications ($CO + 3H_2 \rightarrow CH_4 + H_2O$).

One of the earliest reactions involving the insertion of CO into a C_n-olefin molecule to produce an aldehyde with one greater C_{n+1} carbon number is the so-called hydroformylation or "oxo" reaction. The oxo reaction is carried out over homogeneous catalysts, rhodium or cobalt carbonyls, and is an important industrial process. Recently the production of acetic acid, acetaldehyde, and glycol from CO and H_2 over heterogeneous and homogeneous rhodium catalysts have been reported. Straight-chain saturated hydrocarbons are not the only molecules that may be produced in the Fischer–Tropsch reaction. There have been

Table II. Typical Values of Commercial-Scale Syntheses on Cobalt Catalyst

Constituent	Wt % of Total Products Listed[a]	Olefins (Vol %)	Number of Carbon Atoms	Octane Number, Research Method
Normal-pressure synthesis[b]				
gasol ($C_3 + C_4$)	12	50	$C_3 + C_4$	
gasoline (to 185°C)	49	37	C_4-C_{10}	52
gasoline (to 200°C)	54	34	C_4-C_{11}	49
diesel oil (185°–320°C)	29	15	C_{11}-C_{18}	
diesel oil (200°–320°C)	24	13	C_{12}-C_{19}	
soft paraffins (320°–450°C)	7	iodine value,	$> C_{19}$	
hard paraffins (> 450°C)		2		
Medium-pressure synthesis[c]				
gasol ($C_3 + C_4$)		30	66% C_4 33% C_3	
gasoline (to 185°C)	35	20	C_4-C_{10}	28
gasoline (to 200°C)	40	18	C_4-C_{11}	25
diesel oil (185°–320°C)	35	10	C_{11}-C_{18}	
diesel oil (200°–320°C)	35	8	C_{12}-C_{19}	
soft paraffins (320°C)	30	iodine value,	C_{18}	
soft paraffins (330°C)	25	2	C_{19}	

[a] Total yield per cubic meter of synthesis gas: normal-pressure synthesis, 148 g; medium-pressure synthesis, 145 g of liquid products and 10 g gasol.
[b] At 1 atm; 180°–195°C; catalyst, 100 Co:5 ThO$_2$:7.5 MgO:200 kieselguhr; 1 CO:2H$_2$ (18–20% inert components); throughput 1 m^3 synthesis gas/hr) (kg Co); two stage; no recycle.
[c] At 7 atm, abs; 175°–195°C; catalyst, 100 Co:5 ThO$_2$:7.5 MgO:200 kieselguhr; 1 CO:2H$_2$ (18–20% inert components); throughput 1 m^3 synthesis gas/hr) (kg Co); two stage; no recycle.

early reports of the predominance of isomers among the products when using a promoted thorium oxide, ThO_2, catalyst. This process was called the isosynthesis (1). The typical product distribution yields a large fraction of isobutane. It is interesting to note that elevated temperatures large concentrations of aromatic molecules over the same promoted ThO_2 catalyst are also produced.

Thus thermodynamic considerations and available experimental evidence indicate that by using CO and H_2 mixtures as reactants, one should be able to produce, selectively, a very broad range of hydrocarbon molecules that include alcohols, olefins, acids, and aromatic molecules. Using the proper catalysts, it should be possible to avoid producing the broad product distribution that is found in the conventional Fischer–

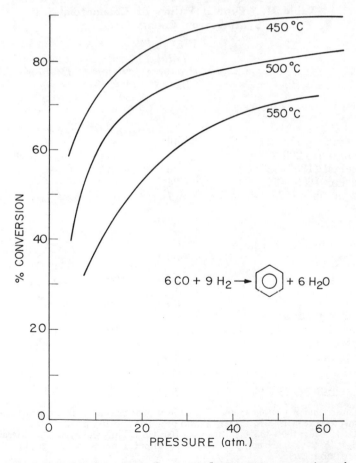

Figure 4. Pressure dependence of the precent conversion of CO and H_2 to benzene. Effect of increasing pressure at different temperatures.

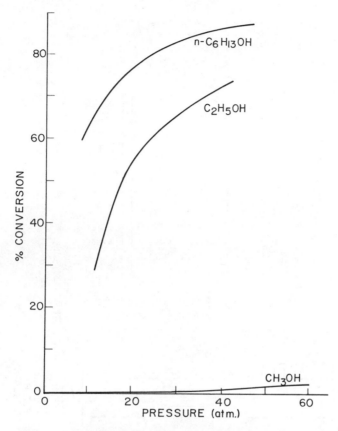

*Figure 5.   Pressure dependence of the precent conversion
of CO and H₂ to alcohols at 400°C*

Tropsch reaction. To this end we must scrutinize the composition and structure of the active catalyst on the atomic scale and change it, if possible, in a way to control the product distribution and the reaction rate.

Our method of investigation is to correlate the reaction rates and product distributions with the catalyst composition and structure. The apparatus we use for this purpose and the experimental procedures are described below.

Experimental

The apparatus as shown in Figure 6 has been described in detail elsewhere (6, 7). It consists of a diffusion-pumped, ultrahigh vacuum bell jar (1×10^{-9} Torr) equipped with a retarding-grid, Auger electron spectroscopy (AES) system, a quadrupole gas analyzer, and a 2-keV ion sputter gun. The unique feature of the apparatus is an internal sample isolation cell which operates as a microbatch reactor (100 cm³ internal

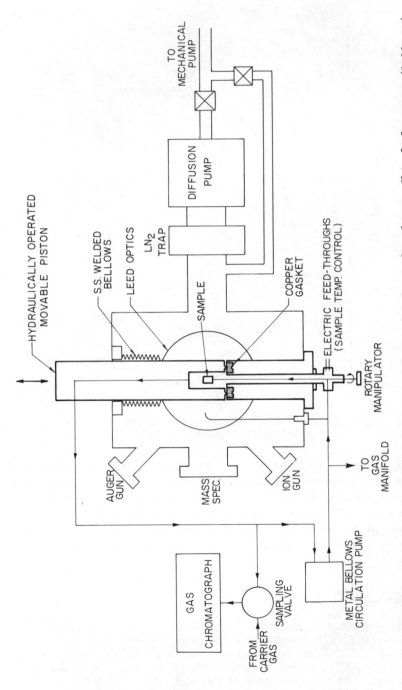

Figure 6. Schematic of UHV surface analysis system equipped with sample isolation cell for high pressure (1–20 atm) catalytic studies

**Table III. Comparison of Polycrystalline Rhodium Foil with a
1% Rh/Al₂O₃ Catalyst in the CO–H₂ Reaction
at Atmospheric Pressure**

	Polycrystalline Rhodium Foil[b] (this work)	*Supported 1% Rh/Al₂O₃[b]*
Reaction conditions	300[b]°C, 3:1 H_2/CO 700 Torr	300°C[a] 3:1 H_2/CO 760 Torr
Type of reactor	batch	flow
Conversion	$< 0.1\%$	$< 5\%$
Product distribution	90% CH_4 ± 3 5% C_2H_4 ± 1 2% C_2H_6 ± 1 3% C_3H_8 ± 1 $< 1\%$ C_4 +	90% CH_4 8% C_2H_6 2% C_3 $< 1\%$ C_4 +
Absolute methanation rate at 300°C (turnover no.)	0.13 ± 0.03 molecules site⁻¹ sec⁻¹	.034 molecules site⁻¹ sec⁻¹
Activation energy (kcal/mol)	24	24

[a] Data adjusted from 275°C.
[b] *See* Ref. 8.

**Table IV. Variation of Reaction-Product Distribution with
H₂/CO Ratio and Temperature, over Rhodium Foils[a]**

Temperature (°C)	Product	H_2/CO = 1/2 (%)	H_2/CO = 3/1 (%)	H_2/CO = 9/1 (%)
250	C_1	65	84	93
	C_2 (=)	16	9	4
	C_2	9.8	3	2
	C_3 +	9.2	4	1
300	C_1	77	89	95
	C_2 (=)	13	7	2
	C_2	4	2	2
	C_3 +	6	3	1
350	C_1	83	94	98
	C_2 (=)	12	3	0
	C_2	1	2	2
	C_3 +	4	1	0.2

[a] C_1 methane; C_2 (=) ethylene; C_2 ethane; C_3 propane.

volume) in the 1–20 atm pressure range while maintaining UHV in the bell jar. An external gas recirculation loop is attached to the cell, through which the reactant gas mixture is admitted. The loop also contains a high-pressure bellows pump for gas circulation and a gas sampling valve that diverts a 0.1-mL sample to a gas chromatograph.

The metal samples were approximately 1-cm² polycrystalline foils (99.99% pure) or single crystals which were pretreated in a hydrogen furnace (1 atm H_2) at 800°C for four days prior to mounting in the vacuum system. This hydrogen treatment was necessary to remove bulk carbon and sulfur impurities which otherwise migrate to the surface during UHV cleaning procedures. The metal samples were mounted such that they could be heated resistively, and the temperature was monitored with a chromel–alumel thermocouple spot welded to the sample edge.

The hydrogen and carbon monoxide used to prepare the synthesis gas were of high purity research grade. The mixtures were prepared in

Table V. Surface Structures of Chemisorbed Small and Pt and on the

Gas	Rh(111) (this paper)	Pd(111)	Ni(111)
H_2	(1×1) or disordered	(1×1)	disordered (2×2)
O_2	(2×2)	(2×2) $(\sqrt{3} \times \sqrt{3})$ R30°	(2×2) $(\sqrt{3} \times \sqrt{3})$ R30°
CO	$(\sqrt{3} \times \sqrt{3})$ R30° split (2×2) (2×2)	$(\sqrt{3} \times \sqrt{3})$ R30° $C(4 \times 2)$ split (2×2)	$(\sqrt{3} \times \sqrt{3})$ R30° $C(4 \times 2)$ $(\sqrt{7}/2 \times \sqrt{7}/2)$ R19.1°
CO_2	$(\sqrt{3} \times \sqrt{3})$ R30° split (2×2) (2×2)	—	(2×2) (2×3)
NO	$C(4 \times 2)$ (2×2)	$C(4 \times 2)$ "star" structure (2×2)	$C(4 \times 2)$ "hexagonal"
C_2H_4	$C(4 \times 2)$	—	(2×2)
C_2H_2	$C(4 \times 2)$	—	(2×2)
C	(8×8) (2×2) R30° $(\sqrt{19} \times \sqrt{19})$ R23.4° $(2\sqrt{3} \times 2\sqrt{3})$ R30° (12×12)	—	(1×1) $(\sqrt{39} \times \sqrt{39})$

the circulation loop, then expanded into the isolation cell. Analysis of the synthesis gas by gas chromatography and mass spectrometry indicated that H_2O in very small amounts was the only impurity.

The clean metal surfaces were prepared in ultrahigh vacuum by ion sputtering (Ar^+, 2 keV, 100 μA) at high temperatures (800°C) for 15–20 minutes, then annealing at 700°C for two minutes. The procedure generally produced a surface that was free from sulfur and oxygen. The only detectable surface impurity after this treatment was carbon (10–15% of a monolayer). Once a clean surface was prepared, the isolation cell was closed and the synthesis gas admitted into cell at the desired pressure. The sample temperature was then raised to 300°C and gas chromatographic sampling of the reaction products was commenced. At any point in the reaction, the cell and circulation loop could be evacuated, the sample cooled to room temperature, and then the cell opened to UHV to allow AES analysis of the surface. The pump-down procedure from 6 atm to 5×10^{-8} Torr took approximately one minute.

Molecules on the (111) Surfaces of Rh, Pd, Ni, Ir, (001) Surface of Ru

Ir(111)	*Pt(111)*	*Ru(001)*
(1 × 1) or disordered	(1 × 1)	—
(2 × 2)	(2 × 2)	(2 × 2)
($\sqrt{3} \times \sqrt{3}$) R30° (2$\sqrt{3} \times$ 2$\sqrt{3}$) R30° split (2$\sqrt{3} \times$ 2$\sqrt{3}$) R30°	($\sqrt{3} \times \sqrt{3}$) R30° C(4 × 2) "hexagonal"	($\sqrt{3} \times \sqrt{3}$) R30° (2 × 2) disorder
		($\sqrt{3} \times \sqrt{3}$) R30° (2 × 2)
—	—	
(2 × 2) [52]		
	—	—
($\sqrt{3} \times \sqrt{3}$) R30°	(2 × 2)	—
($\sqrt{3} \times \sqrt{3}$) R30°	(2 × 2)	—
(9 × 9)	graphite rings	(12 × 12)

Studies of the Hydrogenation of CO and CO₂ over Rhodium

By using the high pressure cell, methanation reaction was studied on initially clean polycrystalline rhodium and iron surfaces. The initial experiments were carried out at 1 atm, where methane is expected to be the main product of the reaction. Table III shows the rate, the activation energy, and the product distribution observed over the small-area rhodium surface and compares these values with those found using dispersed alumina-supported rhodium catalysts under identical conditions (8). The results are in excellent agreement. The identity of these experimental data observed over the surfaces of the same metal but widely different surface structure indicates that methanation is likely to be a structure-insensitive reaction at 1 atm. Changing the H_2/CO ratio does not markedly affect the product distribution under these conditions, as shown in Table IV. Auger electron spectroscopy indicates that during the reaction the active surface is covered with a near monolayer of carbonaceous deposit, but oxygen is not detectable on the surface. The reaction can be interrupted and started up again, the surface remains active indefinitely, and the carbon monolayer appears to reflect the surface composition of the active catalyst at steady state. Oxygen may

Table VI. Surface Structures of Chemisorbed Small

Gas	Rh(100)	Pd(100)	Ni(100)
H_2	(1 × 1) or disordered	—	disordered
O_2	(2 × 2) C(2 × 2)	—	(2 × 2) C(2 × 2)
CO	C(2 × 2) split (2 × 1)	C(4 × 2) R45° compressed C(4 × 2) R45°	C(2 × 2) "hexagonal"
CO_2	C(2 × 2) split (2 × 1)	—	—
NO	C(2 × 2)	—	—
C_2H_4	C(2 × 2)	—	C(2 × 2)
C_2H_2	C(2 × 2)	—	C(2 × 2)
C	C(2 × 2) graphite rings	—	"quasi" C(2 × 2) graphite rings (2 × 2) ($\sqrt{7} \times \sqrt{7}$) R19°

be readily adsorbed on the surface in the absence of H_2 and CO, and it forms ordered surface structures on both (111) and (100) crystal faces of rhodium (9), as shown in Tables V and VI. However, the chemisorbed oxygen is removed rapidly by either CO or by H_2 at temperatures lower (~ 500 K) than those encountered during methanation, as CO_2 or H_2O.

The rapid rate of reaction of oxygen with both CO and H_2 explains the absence of surface oxygen after the reaction. Yet the pretreatment of the surface with oxygen alters the product distribution for a short period (alcohols and other oxygenated products form), and then the methanation reaction becomes predominant again as its steady state is reached. Pretreatment of the surface with C_2H_2 decreases the rate of methanation markedly. The effects of various pretreatments on the rhodium surface for the rate of methanation and the product distribution are summarized in Table VII.

The chemisorption of CO on the clean rhodium and on the metal covered with the carbonaceous deposit show interesting changes. Thermal desorption exhibits only one peak as CO desorbed at a fairly low temperature, 570 K. It appears that the rhodium surface adsorbs and retains molecular CO at 300 K. When CO is adsorbed on the carbon-covered rhodium surface, two thermal desorption peaks appear; one is identical

Molecules on the (100) Surfaces of Rh, Pd, Ni, Ir, and Pt

Ir(100)		Pt(100)	
(1 × 1)	(5 × 1)	(1 × 1)	(5 × 20)
—	(5 × 1) or disordered	(1 × 1) or disordered	(5 × 20) or disordered
(2 × 1)	(2 × 1)	(1 × 1)	not adsorbed
C(2 × 1)	(2 × 2) (1 × 1)	C(2 × 2) (1 × 1)	(1 × 1) C(4 × 2) [46] (2 × 2)
C(2 × 2) (7 × 20)	(2 × 2)	—	—
—	(1 × 1)	—	(1 × 1)
(1 × 1)	(1 × 1)	—	C(2 × 2)
(1 × 1)	(1 × 1)	—	C(2 × 2)
—	C(2 × 2)	—	graphite rings

Table VII. Variation in Methanation Activity, and Product Distributions for the CO–H_2 and CO_2–H_2 Reactions on Clean and Pretreated Rhodium Surfaces[a]

Reaction Gases	Surface[b] Pretreatment	Methanation Rate (300°C) (turnover number)	Product Distribution (%)	
CO–H_2	none	$0.15 \pm .05$	88	C_1
			9	C_2
			3	C_3
CO_2–H_2	none	$0.33 \pm .05$	100	C_1
CO–H_2	O_2	$0.33 \pm .05$	87	C_1
			10	C_2
			3	C_3
CO_2–H_2	O_2	1.7 ± 0.2	98	C_1
			2	C_2
CO–H_2	CO	$0.15 \pm .05$	88	C_1
			9	C_2
			3	C_3
CO_2–H_2	CO	$0.33 \pm .05$	100	C_1
CO–H_2	C_2H_2	$.07 \pm .02$	78	C_1
			18	C_2
			4	C_3
CO_2–H_2	C_2H_2	$.07 \pm .04$	96	C_1
			3	C_2
			1	C_3

[a] Reaction conditions 1:3 ratio, 700 Torr, 300°C.
[b] Heated for 15 min in 700 Torr of the particular gas, then thermally desorbed to 1000°C in vacuo before reaction.

with the peak from clean rhodium and the other is at a much higher temperature, Figure 7. This latter peak can be associated with dissociated CO that partly recombines and desorbs as a molecule only at about 1000 K. These results indicate that adsorbed CO remains in the molecular state on clean rhodium at 300 K but is effectively dissociated at the same temperature on the carbon-covered rhodium surface, implying a drastically different chemical bonding on the two types of surfaces. Carbon monoxide forms a series of surface structures on the (111) and (100) crystal faces of rhodium (9) that indicate a contraction of the surface unit cell as the CO surface coverage is increased at higher ambient pressures.

LEED studies of the chemisorption of CO_2 on the rhodium (111) and rhodium (100) surfaces (9) indicate identical behavior to that of CO (*see* Tables V and VI). The identity of the ordering and structural

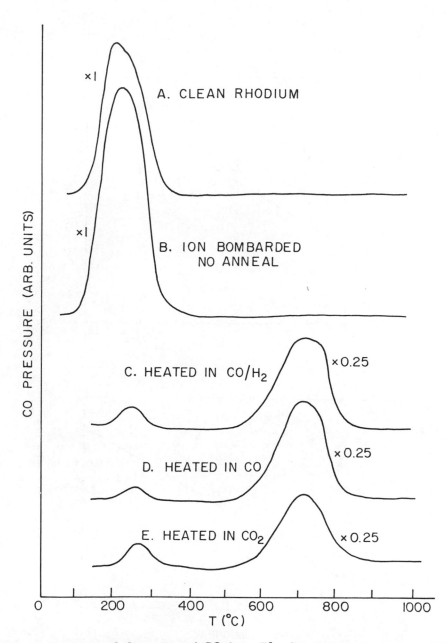

Figure 7.   Thermal desorption of CO from Rh after various treatments; (A) clean surface, 30L; (B) ion-bombarded surface, 30L; (C) heated in CO/H_2 1:1, 10^{-6} Torr, 300°C for 10 min; (D) as in (C) but pure CO; (E) as in (C) but pure CO_2

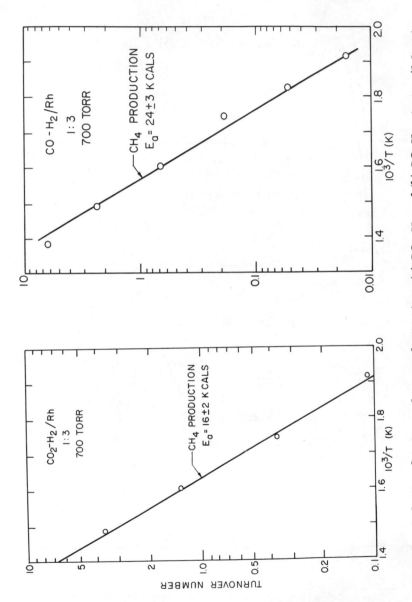

Figure 8.   Arrhenius plots for methane production from (a) CO_2–H_2 and (b) CO–H_2 mixtures (1:3 ratio, 250°–400°C, 700 Torr)

features imply that CO_2 is dissociated on the rhodium surfaces at 300 K to chemisorbed CO and O, and as the oxygen is removed from the surface as O_2, the surface chemistry becomes that of chemisorbed pure CO. The methanation reaction, when it is carried out using CO_2 instead of CO, also clearly reflects this behavior. The CO_2/H_2 mixture produces pure methane, CH_4, and the activation energy for this reaction is 16 kcal/mol as compared with 24 kcal/mol when using CO and H_2 mixtures. These results are shown in Figure 8.

Comparison of Rhodium and Iron for the Hydrogenation of CO and CO₂

Both rhodium and iron polycrystalline foils have been used and compared at 6 atm. Again methanation was predominant even at this pressure. Iron was found to be a better methanation catalyst than rhodium, as indicated by Figure 9. The distribution of higher-molecular-weight products from the two metal surfaces are somewhat different as shown in Figure 10. Iron produces hydrocarbon products up to C_5 under

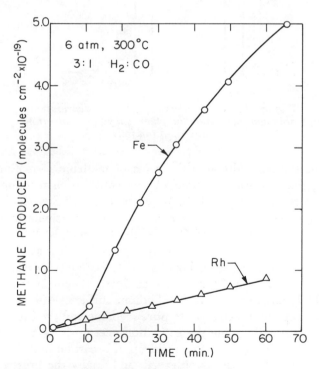

Figure 9.   Total accumulation of methane as a function of reaction time over initially clean rhodium and iron foils

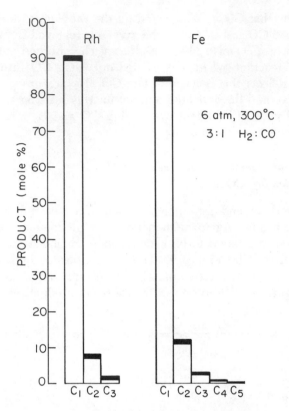

Figure 10. Comparison of product distributions obtained over initially clean polycrystalline iron and rhodium foils

these circumstances. Although the rates of methane formation on iron are higher, the activation energies of the methanation reaction are very similar to those found for rhodium when CO (23 kcal/mol) or CO_2 (15 kcal/mol) were used as reactants. This implies a similarity of surface chemistry for the formation of CH_4 by the two metals. The active iron surface also is covered with a monolayer of carbon just as rhodium was, and one detects no chemisorbed oxygen on the catalyst by Auger electron spectroscopy. However, the iron surface does not remain active for long in the CO/H_2 mixture, unlike the rhodium surface. After 120 minutes, the product distribution changes to pure methane and the rate of reaction slows down markedly while the activation energy drops to 15 kcal/mol. The Auger spectra indicates the build up of a carbon multilayer deposit as this change in reactivity is observed, and finally the presence of iron on the surface can no longer be detected. This is shown in Figure 11. The active iron surface is poisoned rapidly and appears to be unstable under our reaction conditions. Catalyst deactivation is also observed when using CO_2/H_2 mixtures or when the surface is pretreated with oxygen. The initial activity for methanation is higher in these circumstances and the presence of surface oxygen is detectable at first by Auger spectroscopy. However, after a short period of about one hour, the surface oxygen disappears and, shortly, a multilayer carbon deposit forms, effectively poisoning the iron surface.

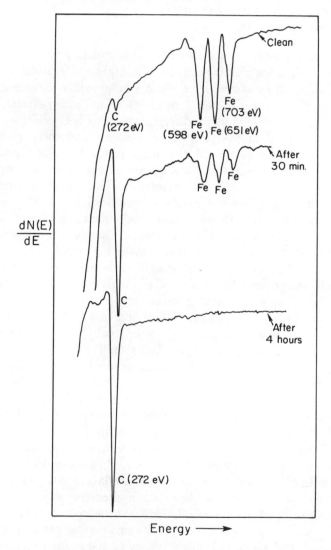

Figure 11. *Auger spectra of the iron surface before and after 30 minutes and after four hours of reaction (6 atm, 3:1 H₂:CO, 300°C)*

By using H_2–Ar mixtures, the multilayer carbon can be removed from the iron surface as methane at a slow rate by the reaction between hydrogen and the surface-carbon multilayers. However, the active iron surface appears to be an unstable methanation catalyst under the same conditions where rhodium was stable. Both active surfaces, however, are covered with a monolayer of carbon.

The Influence of Readsorption on the Product Distribution over Iron

In the present experiments, products with molecular weights greater than that of CH_4 were almost exclusively straight-chain primary olefins. This result is in agreement with previous studies which have suggested that primary olefins are the initial products of a Fischer–Tropsch synthesis (10, 11). However, a major discrepancy exists between the product distribution observed in the present experiments and those previously reported. As shown in Figure 12, the major product in our experiments is methane while in more conventional Fischer–Tropsch studies, the major component is the C_5^+ liquid hydrocarbons. This result is somewhat surprising considering that the reaction conditions are in the range in which Fischer–Tropsch activity should predominate over methanation. One possible explanation for this discrepancy involves the low surface area of our catalyst, which results in extremely low conversions ($< 1\%$). In conventional Fischer–Tropsch studies, high surface area catalysts are used at high conversion. As a result of the high conversion the primary olefins produced initially undergo secondary reactions which result in the formation of paraffins, internal olefins, and methyl-substituted chains (11). It is also known that primary olefins can be incorporated into growing chains or can initiate new chains on the surface (12). This additional reactivity, which does not occur in our low conversion experiments, may play an important role in determining the product distribution.

To investigate the role of readsorption and secondary conversion during Fischer–Tropsch synthesis, experiments were performed in which small amounts of ethylene were added to the synthesis gas before reaction. The fate of the olefin was then followed as a function of reaction time. In the case of ethylene (2.7 mol % in synthesis gas) under the present reaction conditions, 80–90% of the added olefin reacted. As shown in Figure 13, the predominant reaction was hydrogenation to ethane, but approximately 10% of the added ethylene was incorporated into growing chains. The incorporation of ethylene into chain products increased the relative amounts of C_3 to C_5 hydrocarbons as shown in Figure 14. To further demonstrate this effect, a series of experiments were performed in which the initial concentration of ethylene was varied while all other

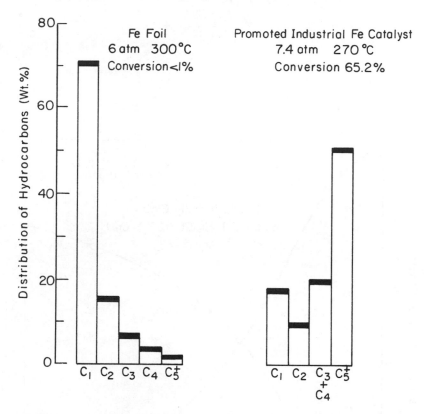

Figure 12. Comparison between the product distribution obtained at low conversion from Ref. 7 with that obtained at high conversions, Ref. 1

reaction parameters were held constant (3:1 H_2:CO, 6 atm, 300°C). Each experiment started with a clean iron surface and the crystal temperature was 300°C. The results of these experiments are summarized in Figure 15 where the product distribution after 90 minutes of reaction is given as a function of initial partial pressure of ethylene. As the initial partial pressure of ethylene increased, the yield of CH_4 remained relatively constant with increasing ethylene in the reactor. The C_5^+ fraction, on the other hand, increased in a linear manner with increasing ethylene partial pressure. The C_5 and C_4 fractions increase to limiting values of 30 and 21 wt%, respectively.

The results of these experiments suggest that the straight-chain primary olefins initially produced over iron catalyst from CO and H_2 can undergo readsorption and secondary reactions. These secondary reactions not only produce the various hydrocarbons associated with the Fischer–Tropsch synthesis but also influence the size of the hydrocarbon chains.

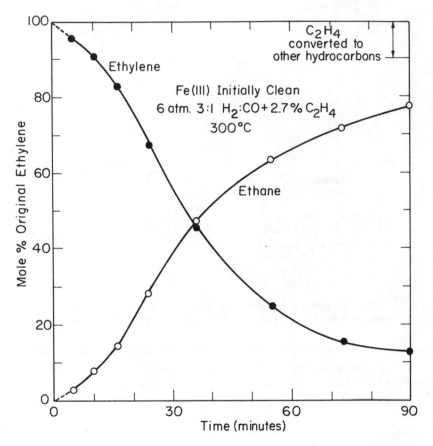

Figure 13.   The conversion of 2.7-mol % added ethylene to ethane as a function of time.   Note that some of the ethylene is converted to other hydrocarbons.

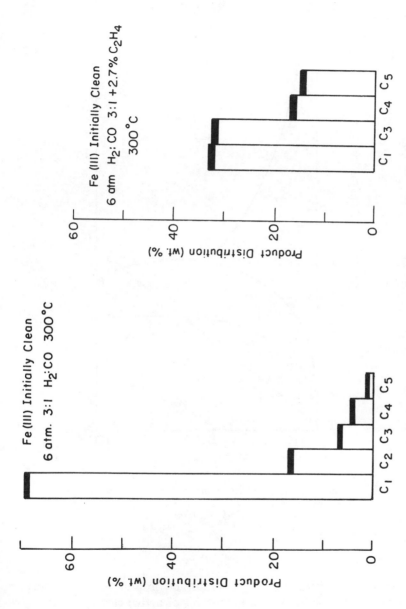

Figure 14. Comparison between the product distribution obtained from initially clean Fe(111) with and without added ethylene. Etheylene concentration is in mol %.

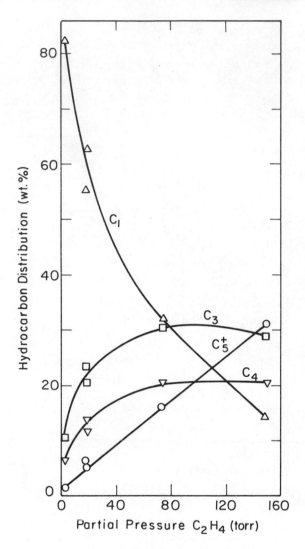

Figure 15. Product distribution for fixed reaction conditions (6 atm, 3:1 H_2:CO, 300°C) as a function of added ethylene

Discussion

Starting with initially clean rhodium and iron surfaces, the surface rapidly becomes covered with a monolayer of active carbon. This active carbon layer appears to hydrogenate directly to produce methane by a mechanism which is very similar for both rhodium and iron. While rhodium–carbon monolayer systems remain stable indefinitely under our experimental conditions, the iron–carbon monolayer surface poisons rapidly and multilayer carbon deposits build up.

A series of interesting studies of the methanation reaction have been carried out by Rabo et al., using pulsed reactor techniques (*13*). Carbon monoxide was adsorbed on supported nickel, cobalt, ruthenium, and palladium catalysts at various temperatures. It was observed that CO adsorbs molecularly and disproportionates into surface carbon and gaseous CO_2 on these metals. The prerequisite for the disproportion reaction ($2CO \rightarrow C + CO_2$) is the weakness of the surface oxygen bond as compared with the carbon–oxygen bond energy in CO_2. Since the appearance of CO_2 is the result of the disproportionation reaction, the amount of carbon on the surface can be titrated by the amount of molecular CO_2 that is produced during CO chemisorption. Using this approach, Rabo et al. determined the fractional surface coverage of both surface carbon and molecular CO; a pulse of H_2 was then admitted at a given surface temperature, and the formation of methane and other hydrocarbon products was monitored. It was found that CO dissociates and yields surface carbon on nickel, cobalt, and ruthenium surfaces at 600 K while it remains molecular on palladium under these conditions. Pulses of H_2 produce CH_4 efficiently by reaction with this surface carbon. The reaction between surface carbon and hydrogen produces methane even at 300 K. When CO was adsorbed on these surfaces at 300 K, it remained largely molecular (as indicated by the absence of CO_2 evolution). The rate of production of CH_4 from adsorbed molecular CO was undetectable above 300 K, but it was detectable and shows at 500 K. It appears that surface carbon produces methane at a higher rate than molecular CO does although the experiment indicates that molecular CO also reacts to form methane at a slow rate. This was proved by the studies which used palladium-catalyst surfaces that produce methane at a slow rate even though only molecular CO is present on the surface. Thus there are two mechanisms of methanation that can be distinguished, involving both active surface carbon and molecular CO. A similar result was obtained by Wise et al. (*14*) who found CH_4 formation from nickel by only one route by the reaction with the surface carbon. Palladium has not only produced methane at a slow rate from adsorbed molecular CO but at 12 atm it also produces methanol very efficiently. Both palladium and platinum exhibit the ability to produce methanol at high pressures. It has been proposed that methanol formation proceeds via the reaction of molecular CO with H_2 while methanation proceeds via the reaction of surface carbon with hydrogen most efficiently.

The nature of the active carbon monolayer on the transition metal surfaces remains to be explored. On heating to 800 K, its activity for methanation is largely lost. Low-energy electron diffraction and Auger electron spectroscopy studies indicate rapid graphitization at these higher temperatures. Thus the bonding of the active carbon (both carbon–carbon and metal–carbon bonds) must be very different from that of graphite. Perhaps carbene-like metal–carbon bonds are responsible for its hydrogen activity. Through the use of electron spectroscopy and vibrational spectroscopy, the properties of this active metal–carbon surface will surely be explored in the near future.

There is a great deal of insight gained into the mechanism of methanation by these and other investigations. There are major differences, however, in the product distribution found on initially clean rhodium and iron catalyst surfaces and on industrial Fischer–Tropsch rhodium and iron catalysts. While rhodium produces oxygenated products and higher-molecular-weight hydrocarbons on the industrial catalysts, it appears to be a poor but stable methanation catalyst by our investigations. Iron produces larger molecular-weight, straight chain hydrocarbons and alcohols industrially while it is an unstable methanation catalyst by our studies. It appears that promoters must play an important role in establishing the product distribution. Potassium, calcium, and manganese are all used as additives to prepare the active and stable catalyst. It appears that the oxygen surface concentration is one of the important ingredients in controlling the product distribution. Not only will it provide oxygen atoms to be built into the forming hydrocarbon molecules, it also can effectively remove part of the surface carbon, thereby reducing the rate of methanation. Oxygen, however, cannot be added from the gas phase as its reactions with both CO and H_2 are rapid and complete. It must be supplied by adding promoters that want strong metal–oxygen bonds or by forming ternary oxide compounds with part of the metal catalyst. By providing oxygen to the metal by surface diffusion, the surface concentration of oxygen can be controlled effectively. One may write a series of reaction steps (15) which express the findings of methanation and methanol formation studies that both molecular CO and surface carbon that form by CO dissociation can be hydrogenated to yield the observed products.

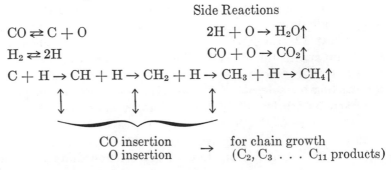

Side Reactions

$$CO \rightleftarrows C + O \qquad\qquad 2H + O \rightarrow H_2O\uparrow$$
$$H_2 \rightleftarrows 2H \qquad\qquad CO + O \rightarrow CO_2\uparrow$$
$$C + H \rightarrow CH + H \rightarrow CH_2 + H \rightarrow CH_3 + H \rightarrow CH_4\uparrow$$

CO insertion → for chain growth
O insertion   (C_2, C_3 . . . C_{11} products)

It can be seen that by changing the oxygen surface concentration the equilibrium of the various surface intermediates may be shifted, which also can change the product distribution in addition to the possibility of incorporating oxygen in the product molecules and removing the surface carbon by oxygen.

There is much less known about the CO insertion reaction which must be an important step in producing high-molecular-weight hydrocarbons. There is good evidence for CO insertion into the metal-olefin bond using the oxo reaction by the following mechanism:

$$HM(CO)_4 \rightleftarrows [HM(CO)_3] + CO$$
$$RCH = CH_2 + [HM(CO)_3] \rightleftarrows RCH = CH_2$$
$$\downarrow$$
$$HM(CO)_3$$
$$+CO$$
$$RCH = CH_2 \rightleftarrows RCH_2CH_2M(CO)_3 \rightleftarrows RCH_2CH_2M(CO)_4$$
$$-CO$$
$$RCH_2CH_2M(CO)_4 \rightleftarrows RCH_2CH_2COM(CO)_3$$
$$H$$
$$RCH_2CH_2COM(CO)_3 + H_2 \rightleftarrows RCH_2CH_2COM(CO)_3$$
$$H$$
$$\downarrow$$
$$RCH_2CH_2CHO + [HM(CO)_3]$$

Assuming that olefin intermediates are produced on the surface of the Fischer–Tropsch catalyst, similar mechanisms for chain growth have been suggested. Alcoholic intermediates also have been proposed based on the isotope-labeling studies of Emmett et al. on iron surfaces. The understanding and control of the insertion reaction appears to be the key for controlling the product distribution. Clearly the mechanism of this reaction will be subjected to close scrutiny in the near future.

It should be noted that both the surface temperature and the pressure are important variables in this reaction that should be studied independently. Much of the carbon monoxide adsorbed on the catalyst surface at 300 K remain molecular while dissociation commences as the temperature is increased. By changing the surface temperature, one can control the ratio of molecular and dissociated carbon monoxide on the surface. Thus, low temperature studies are likely to lead to the formation of higher-molecular-weight hydrocarbons if molecular CO is necessary for the insertion reaction to proceed and if carbon is necessary for the methanation reaction to take place. LEED studies indicated that at high pressure the carbon monoxide surface structure undergoes a contraction of the surface unit cell, indicating an enhanced packing of the CO molecules on the surface. Thus the bonding of the carbon monoxide to the surface changes as a function of pressure. Perhaps the pressure has an important influence on the reaction rate and product distribution because of the changing nature of the surface chemical bond between carbon monoxide and the catalyst surface. Another important variable that may be significant in changing the reaction path is the contact time or residence time of the intermediates and reactants on the surface. The CO hydrogenation reaction is relatively slow, indicating a relatively long residence time on the surface. By arranging the reaction conditions so that the contact time is made shorter or longer, the product distribution may be altered. By controlling the contact time, equilibrium among the various surface intermediates may be established or prevented. Thus the reaction mechanism may be markedly changed as a function of residence time control of the different surface intermediates.

Acknowledgment

The work was supported by the Division of Basic Energy Sciences, U.S. Department of Energy.

Literature Cited

1. Storch, J. H., Golumbic, N., Anderson, R. B., "The Fischer Tropsch and Related Syntheses," Wiley, New York, 1957.
2. Pichler, H., *Adv. Catal.* (1952) **4**, 271.
3. Fischer, F., Tropsch, H., *Brennst. Chem.* (1926) **7**, 97.
4. Klier, K., "Methanol and Methyl Fuels from Syngas," paper presented at University Contractors Conference ERDA, NSF, RANN, EPRI, Park City, UT, 1975.
5. Poutsma, M. L., Elek, L. F., Ibarbia, P. A., Rabo, J. A., unpublished data.
6. Sexton, B. A., Somorjai, G. A., *J. Catal.* (1977) **46**, 167.
7. Dwyer, D. J., Somorjai, G. A., *J. Catal.* (1978) **52**, 291.
8. Vannice, M. A., *J. Catal.* (1975) **37**, 449.
9. Castner, D. G., Sexton, B. A., Somorjai, G. A., *Surf. Sci.* (1978) **71**, 519.
10. Weitkamp, A. W., Seelig, H. S., Bowman, N. J., Cady, W. E., *Ind. Eng. Chem.* (1953) **45**, 344.
11. Olivé, G. H., Olivé, S., *Angew Chem. Int. Ed. Engl.* (1976) **15**, 136.
12. Eidus, Y. I., *Russ. Chem. Rev., Engl. Transl.* (1967) **36**, 339.
13. Rabo, J. A., Risch, A. P., Poutsma, G. A., unpublished data.
14. Wenticek, P. R., Wood, B. J., Wise, H., *J. Catal.* (1976) **43**, 363.
15. Ellgen, P. C., Bartley, W. J., Bhasin, M. M., Wilson, J. P., unpublished data.

RECEIVED July 21, 1978.

Effect of Sulfur on the Fischer–Tropsch Synthesis

Alkali-Promoted Precipitated Cobalt-Based Catalyst[1]

R. J. MADON and W. F. TAYLOR

Exxon Research and Engineering Co., Linden, NJ 07036

Condensed [C_5^+] hydrocarbon product distributions, indicating the occurrence of more than one maximum, were obtained with CO hydrogenation over a precipitated, alkalized $Co:ThO_2:kieselguhr$ catalyst, in the absence and presence of small amounts of sulfur. The sulfur, present in the fixed catalyst bed in the form of longitudinal concentration gradient, tended to increase the molecular weight of the condensed products at reaction pressures of 0.6 and 1.1 MPa. However, at 1.6 MPa, there was no such effect. As bimodal distributions cannot be accounted for by a simple polymerization type of growth mechanism alone, site heterogeneity on promoted catalysts and the occurrence of secondary reactions were invoked as possible factors that could influence hydrocarbon chain growth. Finally, high, identical conversions of H_2 and CO were obtained at several experimental conditions with sulfided and unsulfided catalysts.

Feed gases, CO and H_2, for the Fischer–Tropsch synthesis obtained from coal gasification contain sulfur compounds that have been acknowledged as catalyst poisons. Since the early work of Fischer and Tropsch (1), the need for scrupulous removal of sulfur compounds from reactant gas streams has been stressed. Hence to date, little work has

---

[1] This is the first paper of a series.

0-8412-0453-5/79/33-178-093$05.00/0
© 1979 American Chemical Society

been attempted to investigate the effects of sulfur on Fischer–Tropsch catalysts. Studies of the interaction of poisons on catalysts may be important, and in fact small amounts of poison on catalysts may act beneficially by enhancing catalytic selectivity. This concept of selective poisoning, especially regarding Fischer–Tropsch catalysis, has been reviewed recently (2).

In an early British Patent (3), Myddleton suggested that the primary Fischer–Tropsch products were monoolefins, which were normally easily hydrogenated to paraffins on the active hydrogenation sites of a metallic Fischer–Tropsch catalyst. Myddleton proposed that such hydrogenation sites could be poisoned by small amounts of organic sulfur or H_2S, resulting in enhanced olefin selectivity. In fact, it was found that a $Co:ThO_2$:kieselguhr catalyst promoted with K_2CO_3 could be used to react synthesis gas containing up to 57 mg of organic S/m^3 of gas without any substantial catalyst deactivation, with enhanced olefinic content of the product and increased capacity for the production of heavier condensed hydrocarbons. King (4) claimed that a 100 Co:18 ThO_2:100 kieselguhr catalyst, without alkali promoter, could be used effectively in a synthesis gas stream containing CS_2. And the catalyst containing up to 1.5 wt% S would continue to synthesize liquid hydrocarbons efficiently if the reaction temperature was increased from 185° to 210°C. It should be noted that an unsulfided cobalt catalyst at 210°C would give large quantities of CH_4 and light hydrocarbons. Such observations were further substantiated by Herington and Woodward (5) who also showed that on a similar unalkalized catalyst, small incremental additions of H_2S or CS_2 caused a marked increase in the yield of liquid hydrocarbon product and a decrease in gaseous hydrocarbons.

In this chapter we will discuss efforts at extending the early work. The catalyst used by us was similar to the alkalized cobalt-based catalyst described by Myddleton (3). Our objective was to ascertain, by obtaining detailed analysis of products, how the presence of sulfur could change hydrocarbon selectivity. Furthermore, we wished to study the effect of sulfur on the production of gaseous olefins.

Experimental

Apparatus. The experimental unit, shown in Figure 1, was made of stainless steel and was used for the simultaneous operation of up to seven Alonized (a process which coats the stainless steel with aluminum and alumina; it renders the reactor walls inert for Fischer–Tropsch reactions and, more importantly, for interaction by sulfur compounds) stainless steel reactors, all at the same temperature, pressure, $H_2:CO$ ratio, and space velocity. The unit could be used at pressures up to 3 MPa.

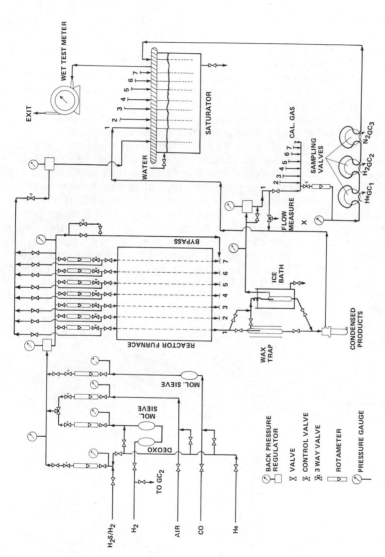

Figure 1. Fischer–Tropsch catalyst test unit

Hydrogen (Linde, 99.99% purity) was passed through a Deoxo unit to remove traces of oxygen and then through a 4A-molecular-sieve trap to eliminate water vapor. CO (Matheson, ultrahigh purity) was also passed through a 4A-molecular-sieve trap. The H_2S/H_2 mixed gas was custom prepared by Matheson and was used without going through any purification steps for sulfiding the catalyst. Finally, provisions also were made to admit helium and air into the unit. The procedure for setting and measuring equal gas flow rates into each reactor has been detailed elsewhere (6). The gas flow rate out of each reactor was obtained either at point X with a soap bubblemeter or with the wet-test meter. As shown in Figure 1, the condensible products from each reactor were recovered in two stages. Waxes and heavy hydrocarbons were collected in the first stage whereas the lighter products were trapped in knockouts immersed in an ice bath. Residual gases were separated and sent to the saturator. When required, these gases were analyzed chromatographically.

The reactor furnace (Figure 2) consisted of an Alonized (alonizing in this case prevents oxidation of the copper) copper pipe, and each reactor was placed in one of the seven equidistant slots provided around the pipe. Three electrical heaters were stacked in the hollow center of the pipe, and the temperature of each heater was controlled separately with a Barber–Colman Model 20 solid-state controller. Besides placing thermocouples in the copper pipe, one thermocouple was embedded in the catalyst in each reactor to record its temperature with the help of a potentiometer and a multi-point recorder. The operation of several reactors in parallel enabled us to evaluate simultaneously the performance of identical catalysts with different amounts of sulfur.

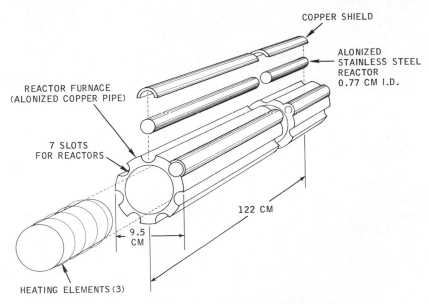

Figure 2.   A tubular, packed-bed multiple reactor system

Analytical Procedures. Exit gases from each reactor were analyzed by using three gas chromatographs. A Hewlett–Packard F and M 720 chromatograph with a thermal conductivity detector and a 3m Poropak Q column was used with hydrogen as a carrier gas to detect CO and CO_2. A similar instrument was used with a 1.8-m 5A-molecular-sieve column and N_2 carrier gas to detect hydrogen. Methane to C_4 gaseous hydrocarbons were analyzed by a Perkin–Elmer 3920 chromatograph equipped with a flame ionization detector and a 3-m *n*-octane on Poracil C column used isothermally at 40°C; helium was used as the carrier gas. The same instrument was also equipped with a flame photometric detector and a 1-m Triton X plus a 2-m Carbopak column to detect small quantities of sulfur compounds.

The carbon number distribution of the condensed products was obtained with a Perkin–Elmer 900 chromatograph, using a flame ionization detector and a 3-m column containing 10% SP 2100 on 80–100 mesh Supelcoport. The column was temperature programmed from 60° to 350°C at 8°C/min and held at 350°C. Helium, flowing at 30 cm^3/min, was used as the carrier gas.

The weight percent of sulfur on the catalysts was obtained via the high-temperature-combustion Dietert technique, ASTM method D1552. The sulfur analysis was based on the weight of the catalyst after all carbon had been removed, i.e., the weight percent reported is not masked by any residual wax or coke on the spent catalyst.

Catalyst. The catalyst, 100 Co:16 ThO_2:93 kieselguhr:2 K_2CO_3, was specially prepared by Harshaw Chemical Company. Johns–Manville Celite, FC Grade, was acid treated by digesting it for 5 hr at 70°–80°C with a nitric acid solution. After washing with deionized water, the material was calcined at 650°C for 2 hr. This material was used as the substrate. A boiling soda-ash solution and a boiling kieselguhr slurry were added simultaneously to a boiling Co–Th nitrate solution, and the pH was adjusted between 7.5–8.0. After filtering, the cake was washed with deionized water at room temperature. The cake was dried at 110°C and treated with K_2CO_3 in a water–alcohol solution. The K_2CO_3 was based on the available cobalt by analysis. After drying, the catalyst was mixed with 0.5% graphite. Slugs 1.3 cm in length were made with a RB-2 tabletting machine and subsequently were crushed and sized to a 60–120 mesh powder.

The reactors were filled with 50 cm^3, 29.8 g, of the catalyst, and after being assembled on the unit, they were first flushed and then pressure tested with helium. The pressure was then reduced to atmospheric pressure, and the temperature of the catalyst was then raised to 400°C under flowing hydrogen at a gas hourly space velocity (GHSV) of 350–375 v/v/hr. The catalyst was then reduced for 4 hr in flowing hydrogen at atmospheric pressure and 400°C. The catalyst temperature was dropped to 140°C in flowing hydrogen. The catalysts in all the reactors were then treated at atmospheric pressure with flowing H_2 and CO, H_2: CO = 2, at a space velocity of 300 v/v/hr; first at 140°C for 1 hr, at 150°C for 1 hr, at 165°C for 13 hr, at 170°C for 8 hr, and finally at 180°C for 18 hr. The reactors were finally flushed with helium before sulfiding. This pretreatment was similar to the one used for cobalt-based catalysts in the past (5, 7, 8) and suggested by Herington and Woodward (5) as being useful for producing liquid hydrocarbons immediately on cobalt catalysts without an extended induction period in which methane would be the principal initial product. The effect of varying pretreatment on catalyst performance was not studied here.

After each experiment the catalyst was kept overnight in flowing hydrogen at the same temperature and pressure used in the experiment. The flow of hydrogen was stopped just before the next experiment was started. This catalyst conditioning has been recommended for rejuvenating cobalt catalysts (5), and it was used by us to prevent the catalyst from undergoing drastic changes during the course of the run. The reproducibility of the experimental results was good and has been discussed in detail elsewhere (6). However, reactor 3 plugged at various times during experimentation; hence results from this reactor will be looked at only briefly.

The catalysts were sulfided at 180°C and atmospheric pressure with a 2:1 mixture of H_2:CO containing 250 ppm H_2S flowing at a space velocity of 300 v/v/hr. Sulfiding periods were varied for the different reactors depending on the nominal weight percent of sulfur that was required. No H_2S was detected coming out of the reactors during the sulfiding process. After the required amount of sulfiding was completed, H_2S was not added again to the catalyst at any time during experimentation; all experiments were carried out with pure CO and H_2. The nominal

Table I. Sulfur Distribution in Catalyst Beds

	Reactor 1	Reactor 2	Reactor 3	Reactor 4
Nominal S level				
as wt% of unreduced catalyst	0.29	0.43	0.50	0.50
as mg S/g Co	8.2	12.2	14.2	14.2
Longitudinal S distribution (% S by wt)				
Section				
1	4.96	5.36	7.66	7.57
2	1.32	2.07	2.53	2.17
3	1.02	0.81	1.38	0.74
4	0.06	0.09	0.05	0.05
5	0.05	0.04	0.08	0.06
6	0.03	0.46	0.01	0.02
7	0.03	0.16	0.04	0.05
8	0.01	0.13	0.02	0.03
9	0.02	0.08	0.04	0.03
10	0.03	0.07	0.04	0.05
11	0.01	0.10	0.05	0.01
12	0.02	0.09	0.02	0.02
13	0.01	0.08	0.02	0.05
14	0.01	0.10	0.03	0.03
15	0.01	0.03	0.02	0.01
16	0.01	0.08	0.01	0.02
17	0.01	0.04	0.02	0.06

sulfur levels calculated as weight percent of unreduced catalyst and as milligrams of sulfur per grams of cobalt are shown in Table I. However, as nominal sulfur levels do not give complete information regarding the amount of sulfur on the catalyst, a detailed longitudinal sulfur gradient analysis was performed (Table I). At the completion of the run, the catalyst to be analyzed was removed from the reactor in equal sections. Each section corresponded approximately to 5–6 cm of reactor length. In Table I, section 1 corresponds to the first 5–6 cm of the inlet side of the reactor, section 2 corresponds to the next 5–6 cm of reactor length, etc.

In all cases the inlet portion, first 20 percent, of the bed contains most of the sulfur. The amount of sulfur after the first 20 percent of bed length is greater in reactor 2 than in the other reactors, i.e., sulfur distribution is a little better in reactor 2; therefore it will be especially important to compare the results obtained from reactor 2 with other results in the run. Sulfur concentration in reactor 5 was negligibly small throughout the bed, nominal sulfur content being $< 0.01 \%$ S by weight.

Results and Discussion

As the intrusion of transport phenomena can mask significantly the intrinsic behavior of catalysts, and as this problem is more pronounced during integral conversions, a detailed analysis of various transport artifacts was completed and is discussed elsewhere (6). At the experimental conditions reported here, it was shown (6) that axial temperature gradients did not exist. The most problematic artifact, interparticle heat transport, was minimized in our case by using long, narrow reactors; a reactor of internal diameter not greater than 0.8 cm is crucial for avoiding radial temperature gradients. Previous research by Roelen at Ruhrchemie (9) showed that the rate on a cobalt-based catalyst was independent of pressure between 20–101 kPa. The same result was found at USBM (10, 11) between 101–1500 kPa. As the apparent activation energy obtained by others (11) was approximately 84 kJ mol^{-1}, external diffusion was not controlling the reaction. Indeed, if the true order of reaction is zero, the problems of external diffusion would not be present, and therefore external mass transport was not influencing our observations. Also, as shown elsewhere (6), by using catalyst particle size less than 0.02 cm, problems associated with pore diffusion were eliminated.

Using a point rate, an approximate turnover frequency (N) can be obtained for the reaction on our cobalt catalyst. The turnover frequency is defined as the moles of a reactant consumed per surface mole of active material per second. A cobalt-metal particle size of 11.5 nm was obtained by x-ray diffraction measurements on the used catalyst. This metal particle size corresponds to a metal dispersion (D), i.e., the fraction of surface cobalt atoms, of approximately 0.08. Assuming that after catalyst reduction all the cobalt on the support occurs as the metal, the value of N_{CO} obtained for the unsulfided catalyst is 2.0×10^{-3} sec^{-1} at 197°C, 0.6

MPa total pressure, and H_2:CO equal to 1.9. This may be a conservatively low value, as all the cobalt on the support may not be metallic. The value of N_{CO} for methanation at atmospheric pressure obtained by Vannice (12) on a 2% Co/Al_2O_3 catalyst, $D = 0.08$, was 4.6×10^{-4} sec^{-1} after extrapolation to 197°C. Furthermore, Vannice (12) obtained a $- 0.5$ order dependence on CO pressure. The two values of N_{CO} for FT and methanation seem to be close. However, when compared with other reactions such as hydrogenation of olefins, where N at ambient conditions is about 1 sec^{-1} (13), the Fischer–Tropsch reaction can be seen to be quite slow.

Table II compares the activity, stated as a percent conversion, and the selectivity with C_5^+ hydrocarbons. The conversions during all experiments were high, the lowest observed conversions occurring when 1.5 H_2:CO ratio was used. On comparing sulfided and unsulfided catalysts, it can be seen that under certain experimental conditions, the sulfided catalysts, especially the one with 0.43 wt% S, gave conversions 20–30% below those obtained for the unsulfided catalyst. This is a reasonable observation, as about 20% of the catalyst bed has a significant amount of sulfur. However, under certain other experimental conditions, e.g., when the H_2:CO ratio is 1.5, the conversions obtained from all reactors were close; no significant deactivation was observed. One argument may be that conversion values from all reactors are close because the complete bed does not participate during the reaction, and therefore even though the sulfided portion of the bed remains inactive, the rest of the catalyst is sufficient to give the appropriate conversion. This argument is valid when conversions of 99–100% are obtained. But, when conversions of about 70–90% are obtained, the above argument cannot explain the similarity of results on the sulfided and unsulfided catalyst.

Table II. Conversion of Carbon Monoxide and Hydrogen

S Nominal Wt %: *Expt. Cond.*[a]		*Total (H_2 + CO) Conversion (%)*			
		0.29	*0.43*	*0.5*	*0*
H_2:CO	*Pressure (MPa)*				
1.9	0.6	86 (81)[b]	75 (85)	73 (85)	91 (84)
1.5	0.6	82 (83)	72 (82)	78 (85)	80 (83)
2.1[c]	0.6	91 (82)	86 (74)	90 (74)	87 (82)
1.9	1.1	98 (79)	92 (80)	98 (75)	96 (78)
1.5	1.1	83 (82)	79 (86)	79 (78)	81 (80)
2.0	1.6	97 (84)	67 (70)	88 (83)	87 (75)

[a] Temperature $= 197° \pm 3$°C, GHSV $\simeq$ 200–210 v/v/hr.
[b] ( ) = Selectivity, % CO converted to C_5^+ hydrocarbon.
[c] GHSV $\simeq$ 250 v/v/hr.

Table III. Olefins in Gaseous Products

S Nominal Wt %: Expt. Cond.[a]		Propylene:Propane				1-Butene:n-Butane			
		0.29	0.43	0.50	0	0.29	0.43	0.50	0
H_2:CO	Pressure (MPa)								
1.9	0.6	1.45	2.22	2.25	1.25	0.79	1.35	1.39	0.64
1.5	0.6	2.33	2.55	2.43	2.37	1.36	1.58	1.44	1.40
1.9	1.1	0.62	1.18	—	0.61	0.19	0.60	—	0.23
1.5	1.1	1.80	2.47	2.00	2.22	1.08	1.55	1.06	1.36
2.0	1.6	0.41	1.59	1.29	0.82	0.13	0.97	0.69	0.39

[a] Temperature = 197° ± 3°C, GHSV ≃ 200–210 v/v/hr. For all conditions, ethylene:ethane ratios were negligible.

This is an important result as it shows that under certain conditions, possibly at low H_2:CO ratios, a cobalt-based catalyst, such as the one used here, can withstand deactivation by sulfur to a certain extent.

The selectivity values given in parentheses in Table II are quite uniform and seem to be independent of the presence of sulfur at the reaction conditions used by us. Unlike the observations of Herington and Woodward (5), we did not observe any drastic improvement in liquid hydrocarbon selectivity or decrease in gaseous products. The selectivity to C_5^+ hydrocarbons was consistently high, approximately 75–85%, indicating that less than 25% of the CO was used to make gaseous hydrocarbons and CO_2.

It has been shown (14, 15) that the primary products of the Fischer–Tropsch reaction are α-olefins which may be hydrogenated in a consecutive step to give the corresponding paraffins. If this is the case, then it is interesting to note, by comparing olefin:paraffin ratios (Table III), how sulfur influences the hydrogenation capacity of the catalyst. For all conditions, the catalyst in which the sulfur was relatively well distributed, reactor 2—0.43 wt % S, consistently gave olefin:paraffin values greater than those obtained from the unsulfided catalyst. High values were also obtained in several experiments for the catalyst containing the most sulfur, 0.5 wt% S, reactor 4. This means that under certain conditions sulfur can interact with the catalyst surface to reduce its effectiveness for olefin hydrogenation. The effect of sulfur was significant only when the higher H_2:CO ratio of 1.9 was used; the sulfur effect was minimal at the lower H_2:CO ratio of 1.5. Variation of the olefin:paraffin ratio also depended on the process pressure, and it was highest at 0.6 MPa when compared with other pressures at a particular value of H_2:CO. Although it is difficult to explain, a more efficient hydrogenation took place at the intermediate pressure of 1.1 MPa and a H_2:CO value of 1.9 rather than at

1.6 MPa. Once again, however, as in the case of the sulfur effect, the effect of pressure for varying the olefin:paraffin ratio was important only at the high, 1.9, H_2:CO ratio. Thus, operation at a low H_2:CO ratio not only gives the highest olefin:paraffin ratio but also precludes any significant sulfur or pressure effect. It is important to note here how different process conditions can influence the way in which sulfur affects catalyst performance.

Though, as seen above, a definite effect of sulfur exists for retarding the hydrogenation of gaseous olefins, a more salient observation is the effect of sulfur on the distribution of condensed hydrocarbons. These results are shown in Figures 3, 4, 5, and 6.

Figures 3a and 3b refer to results obtained at our lowest operating pressure of 0.6 MPa. Product distributions from the unsulfided catalyst, reactor 5, and from the catalysts containing 0.29 wt% S, reactor 1, and 0.43 wt% S, reactor 2, show a similar trend with one maximum at C_{13}; a distinct shoulder S_1 for reactor 5 and a slight shoulder for reactors 1 and 2 are present at about C_{24}. The results from reactor 4, Figure 3b, which contains the largest amount of sulfur, are quite different, showing a bimodal distribution with one maximum at C_{13}, like the other reactors, a minimum at C_{19}, and a second maximum at C_{23}. It appears that the shoulder S_1 of the curve in Figure 3a has grown in Figure 3b to give the second maximum at C_{23}; this increase in higher-molecular-weight products may be attributed to the larger quantity of sulfur in reactor 4.

Figures 4a and 4b refer to the results obtained at the next higher operating pressure of 1.1 MPa. Reactors 1, 2, and 5 once again show similar results, but here they all give a bimodal distribution with the first maximum at C_{10}–C_{13}, a minimum at C_{16}–C_{18}, and a second maximum at C_{21}–C_{22}. It is interesting to note that the minimum in the case of reactor 2, 0.43 wt% S, is not as sharp as that obtained with reactors 1 and 5. And as the sulfur content increases to 0.5 wt%, this minimum disappears and only one maximum at C_{21} is observed. The first maximum in Figure 4a seems to appear as shoulders S_2 and S_3 in Figure 4b. Once again, it seems that the products of reactors 3 and 4 are heavier, i.e., the maximum at C_{21} has grown perhaps at the expense of the first maximum of reactors 1, 2, and 5. Furthermore, products up to C_{41} are obtained with catalysts containing 0.5 wt% S and also 0.43 wt% S, whereas the unsulfided catalyst only gives products up to C_{31}.

Finally, Figure 5 refers to results obtained at our highest operating pressure, 1.6 MPa. Here all of the reactors give an almost identical trend including the small but definite shoulder S_4 at about C_{25}. The initial maximum seems to be split at C_{11} and C_{18}; this is most evident for reactor 2 and differs only slightly for each reactor. The important fact is that at the highest pressure, similar amounts of heavy hydrocarbons

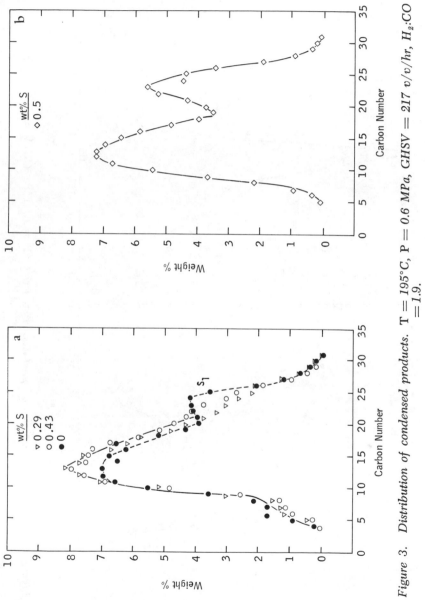

Figure 3. Distribution of condensed products. T = 195°C, P = 0.6 MPa, GHSV = 217 v/v/hr, H₂:CO = 1.9.

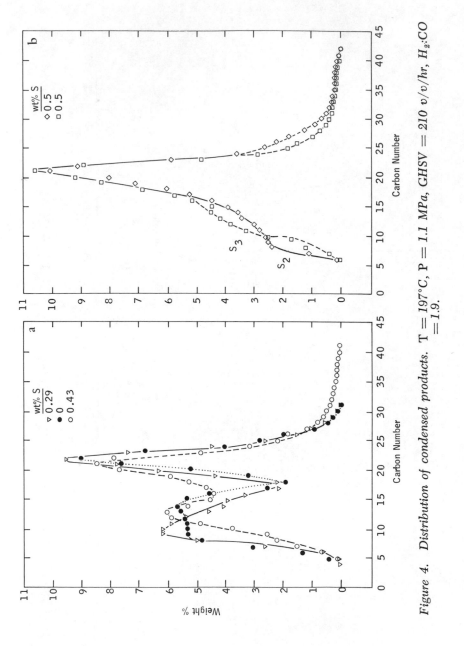

Figure 4.   Distribution of condensed products. T = 197°C, P = 1.1 MPa, GHSV = 210 v/v/hr, H₂:CO = 1.9.

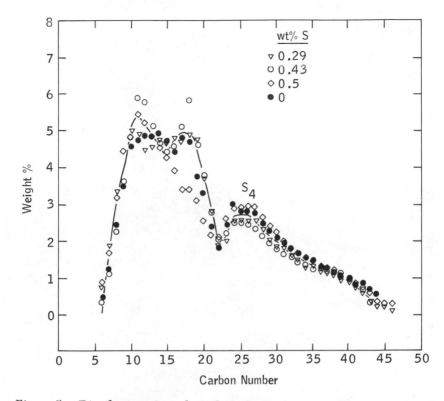

Figure 5.   Distribution of condensed products. T = 197°C, P = 1.6 MPa, GHSV = 209 v/v/hr, H₂:CO = 2.0.

are formed in all cases, and the effect of sulfur on the carbon-number distribution is small; i.e., the pressure effect dominates the sulfur effect. This is quite different from the results at 0.6 and 1.1 MPa.

The above descriptions of the sulfur effect were for results at various pressures but with a high H_2:CO ratio of 1.9. The product distributions (Figures 6a and 6b) from an experiment performed at 1.1 MPa and H_2:CO equal to 1.5 are quite different from those obtained from an experiment performed under the same conditions but where the H_2:CO ratio was 1.9 (Figures 4a and 4b). As shown in Figure 6a, reactors 1 and 5 show similar distributions but with three maxima at C_{10}, C_{17}, and C_{22}–C_{23}, the last maximum being more like a shoulder. Products from reactors 2, 3, and 4 (Figure 6b) show a broad distribution of heavier hydrocarbons with no sharp peaks but with a small shoulder at C_{21}.

A generalized examination of the distribution curves from all reactors in any one experiment shows that reactor 5, 0 wt% S, and reactor 1, 0.29 wt% S, give results showing similar trends but which are different from

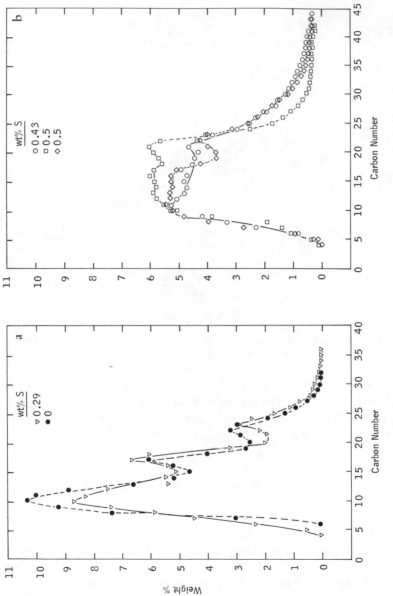

Figure 6.   Distribution of condensed products. T = 195°C, P = 1.1 MPa, GHSV = 200 v/v/hr, H$_2$:CO = 1.5.

those of reactors 3 and 4, 0.5 wt% S. The results from reactor 2, 0.43 wt% S, depending on the experimental conditions, are either like those of reactors 1 or 4. Closer examination of the curves shows that the results of reactor 4 differ because they usually contain heavier hydrocarbons than the products obtained from the unsulfided catalyst. For example, in the bimodal curve in Figure 3b a second peak with a maximum at C_{23} is observed, whereas only a slight shoulder is evident near the same area in Figure 3a. This increase of heavier products, seen in Figure 3b, which leads to a bimodal distribution may be attributable to the effect of sulfur. The same trend is seen in the results from other experiments except for the case when the reaction pressure was 1.6 MPa. At this high pressure the effect of sulfur was negligible. It seems that the product distribution is affected by the presence of sulfur and by the experimental parameters.

Let us summarize the important observations from the results of the product distributions described above. (A) Distributions with more than one maximum have been obtained; (B) a peak or shoulder always exists at approximately C_{20-25}; (C) in experiments conducted at 0.6 and 1.1 MPa, heavier hydrocarbon products are formed when larger amounts of sulfur are present in the catalysts, but at 1.6 MPa the effect of sulfur on the distribution is small; (D) increasing the pressure increases the formation of heavier products; and (E) varying the $H_2:CO$ ratio changes the product distribution.

The mechanism of the Fischer–Tropsch synthesis has been stated to be akin to a polymerization mechanism, and the chain growth may be attributable to the stepwise addition of CH_2 species (16), to surface condensation with oxygenated surface species (11), or to CO-insertion-type reactions (17). In fact, a variant of the Schulz–Flory distribution law for describing molecular weight distributions in polymerization processes has been used in the past by Anderson and co-workers (8, 15) to analyze Fischer–Tropsch hydrocarbon data. The law has recently been used (18) in its original form to fit data on cobalt-based and iron-based catalysts; the fit occurs between C_4 and C_{12}. Such a law precludes the occurrence of more than one maximum, emphasizes the single-carbon stepwise-growth model, and is derived with the assumption that the rates of chain growth and termination are independent of chain length. Recently, Pichler et al. (19) analyzed the change of growth rate of hydrocarbons and concluded that the concept of chain growth in one-carbon steps was insufficient to explain the construction of larger hydrocarbon molecules. The fact that bimodal distributions have been observed by us signifies that besides the primary polymerization-type process, important secondary reactions may occur and affect the growth of hydrocarbon molecules. As only a small amount of products heavier than about C_{10} leave the reactor in the gas phase, their residence time is large, and hence they are likely to be involved in secondary reactions. The more recent articles (*14, 18, 19, 20*) on the Fischer–Tropsch synthesis have stated this point.

Since the pioneering work of Eidus and co-workers (*21, 22*), and Kölbel and Ruschenburg (*23*) on the role of olefins in the Fischer–Tropsch synthesis, recent results (*20*) using C^{14}-tagged olefins have established the fact that olefins participate in the synthesis reaction to a considerable degree. It also has been established that α-olefins are the primary Fischer–Tropsch products (*14*) and that they are consequently hydrogenated to the respective alkane. Alkanes, after being desorbed, do not behave like their precursor alkenes and do not participate in the Fischer–Tropsch reaction (*20*). The C^{14} tracer studies on $Co:ThO_2:$kieselguhr catalyst conducted by Schulz et al. (*20*) contained several important observations. It was found that the overall conversion of tracer olefins was more than 90%. Although hydrogenation to paraffins constituted the principal reaction of olefins, other secondary reactions played an essential role. Olefins can crack, initiate a chain by forming a chemisorption complex which can grow further with CO and H_2, interact with other hydrocarbon chains which are growing on the catalyst—multiple build-in (*24*)—and, especially in the case of ethylene, terminate hydrocarbon chains growing on the catalyst. Perhaps the most important result is that this activity of olefins is not limited to ethylene, propylene, and butene but is also seen for larger molecules such as 1-hexadecene. One intriguing conclusion by Schulz et al. is that 1- and 2-C atoms of 1-hexadecene seem to be transferred to other groupings growing on the catalyst. These reactions of olefins may take place after they have initially desorbed and then readsorbed in another part of the system. In our experiments negligibly small amounts of ethylene were observed. In comparison, relatively large quantities of propylene and 1-butene were obtained; in fact, depending on the experimental conditions, often more of the C_3 and C_4 olefins were formed than their respective paraffins. The rate of hydrogenation of ethylene to ethane is probably not very much greater than the corresponding rates for propylene and 1-butene. Therefore, more ethylene could have been consumed in secondary reactions.

Hall, Kokes, and Emmett (*24*), who used radioactive tracers to study the Fischer–Tropsch reaction, suggested that besides stepwise growth with single-carbon intermediates, multiple build-in could and probably did occur in the synthesis. They also showed that multiple build-in could not be distinguished from single-carbon stepwise growth below C_{12}–C_{16} hydrocarbons, and thus the effect of multiple build-in could only be seen if detailed product distributions up to large carbon numbers were obtained.

The hypothesis of multiple build-in where a chain C_i can interact with a chain C_j leads one to reflect on the possibility of a chain termination by combination. If reactions were occurring in which termination could occur by simple desorption and also by combination, two peaks would be observed. The second maxima would have a center at approximately twice the value of the first, as doubling of the most prevalent adsorbed chain lengths is likely (25). Furthermore, secondary events such as those discussed above or chain transfer could cause the distributions of the two peaks to be different from one another. Thus the fact that secondary reactions during Fischer–Tropsch synthesis occur and that multiple build-in and termination by combination are viable propositions help rationalize distributions that do not follow the Schulz–Flory law and appear with more than a single maximum.

However, besides such mechanistic rationalizations for our observations, it is also necessary to note that surfaces of promoted catalysts such as the one used by us may not be uniform and that site heterogeneity may play a significant role in determining product distributions. The existence of different polymerization sites could give rise to more than one kind of polymerization process, leading to the superposition of separate curves. Our results consistently indicate product distributions that show peaks or shoulders at about C_9–C_{12} and C_{20}–C_{25}. In the absence of sulfur the sites that produce the first peak seem to dominate, whereas sulfur seems to be influencing product distribution by enhancing the role of sites that produce heavier hydrocarbons. Experimental conditions may also influence product distributions by affecting growth on the different sites.

A tentative explanation of our results may be given by invoking both arguments; sulfur predominantly affects growth by influencing polymerization sites on a nonuniform catalytic surface, and experimental conditions affect growth by influencing variations in growth mechanisms such as secondary reactions, multiple build-in, and termination by combination. However, at this stage there is insufficient data to further evaluate the relative importance of the two reasons, site heterogeneity and mechanistic variations, that have been given to explain our observations.

Before concluding, we would like to reiterate the fact that the sulfided catalyst beds contained a severe longitudinal sulfur-concentration gradient, with most of the sulfur covering about 20% of the catalyst bed near the reactor entrance. Yet sulfur effects, subtle in the case of gaseous olefins produced and significant in the case of condensed product distributions, were obtained. These observations indicate that either a very small amount of sulfur is necessary to cause observable changes in cobalt-based catalyst performance and/or that reactions taking part in one section of the fixed-bed reactor may be dependent on reactions that have taken place in sections before it. Further evaluation should be done by comparing catalyst beds with an even distribution of sulfur with those having a sulfur concentration gradient and by performing microcatalytic reactor studies that simulate initial sections of catalyst beds that contain sulfur concentration gradients. The way in which sulfur is added to a catalyst may be quite important and may influence the same catalyst differently.

Acknowledgment

This work was supported by the Department of Energy under Contract No. E(46-1)-8008. The authors are particularly indebted to E. R. Bucker who helped perform the experiments. Valuable contributions were also made by E. Calvin, W. McSweeney, and E. Vath. It is also a pleasure to acknowledge many fruitful discussions with Professors M. Boudart and D. Ollis.

Literature Cited

1. Fischer, F., Tropsch, H., *Brennst. Chem.* (1926) **7**, 97.
2. Madon, R. J., Shaw, H., *Catal. Rev.—Sci. Eng.* (1977) **15**, 69.
3. Myddleton, W. W., British Patent 509,325, 1939.
4. King, J. G., *J. Inst. Fuel* (1938) **11**, 484.
5. Herington, E. F. G., Woodward, L. A., *Trans. Faraday Soc.* (1939) **35**, 958.
6. Madon, R. J., Bucker, E. R., Taylor, W. F., Department of Energy, Final Report, Contract No. E(46-1)-8008, July, 1977.
7. Anderson, R. B., Krieg, A., Seligman, B., O'Neill, W. E., *Ind. Eng. Chem.* (1947) **39**, 1548.
8. Anderson, R. B., "Catalysis," P. H. Emmett, Ed., Vol. IV, Reinhold, New York, 1956.
9. Roelen, O., in Ref. 8.
10. Anderson, R. B., Hall, W. K., Krieg, A., Seligman, B. J., *J. Am. Chem. Soc.* (1949) **71**, 183.
11. Storch, H. H., Golumbic, N., Anderson, R. B., "The Fischer-Tropsch and Related Synthesis," John Wiley and Sons, New York, 1951.
12. Vannice, M. A., *J. Catal.* (1975) **37**, 449.
13. Schlatter, J. C., Boudart, M., *J. Catal.* (1972) **24**, 482.
14. Pichler, H., Schulz, H., Hojabri, F., *Brennst. Chem.* (1964) **45**, 215.
15. Friedel, R. A., Anderson, R. B., *J. Am. Chem. Soc.* (1950) **72**, 1212, 2307.
16. Eidus, Ya. T., *Russ. Chem. Rev. (Eng. Transl.)* (1967) **36**, 338.
17. Pichler, H., Schulz, H., *Chem. Ing. Tech.* (1970) **42**, 1162.
18. Henrici-Olivé, G., Olivé, S., *Angew. Chem., Int. Ed. Engl.* (1976) **15**, 136.
19. Pichler, H., Schulz, H., Elstner, M., *Brennst. Chem.* (1967) **48**, 78.
20. Schulz, H., Rao, B. R., Elstner, M., *Erdoel Kohle* (1970) **23**, 651.
21. Eidus, Ya. T., Zelinskii, N. D., Puzitski, K. V., *Bull. Acad. Sci. U.S.S.R., Cl. Sci. Chim.* (1949) 110.
22. Eidus, Ya. T., Ershov, N. I., Batuev, M. I., Zelinskii, N. D., *Bull. Acad. Sci. U.S.S.R., Cl. Sci. Chim.* (1951) 722.

23. Kölbel, H., Ruschenburg, E., *Brennst. Chem.* (1954) **35**, 161.
24. Hall, W. K., Kokes, R. J., Emmett, P. H., *J. Am. Chem. Soc.* (1960) **82**, 1027.
25. Ollis, D., private communication.

RECEIVED June 22, 1978.

Chain Growth in the Fischer–Tropsch Synthesis

A Computer Simulation of a Catalytic Process

ROBERT B. ANDERSON and YUN-CHEUNG CHAN[1]

Department of Chemical Engineering and Institute of Materials Research, McMaster University, Hamilton, Ontario

Recent detailed analyses of Fischer–Tropsch hydrocarbons by Pichler and Schulz have been used to examine the growth of the carbon chain. The presence of ethyl-substituted isomers requires modification of the simple chain growth scheme involving one-carbon addition to the first or second carbon at one end of the growing chain. With two constant-growth parameters, independent of chain length and structure, the isomer distributions from iron and cobalt could be accurately predicted by four of the nine schemes tried. The chain-growth process was simulated using a digital computer. The carbon chain was represented by an array of numbers or vector with the number 1 denoting a carbon atom; 2, a methyl-substituted carbon; and 3, an ethyl-substituted carbon.

The products of the Fischer–Tropsch synthesis are generally not in thermodynamic equilibrium with each other or with the reactants $(1, 2, 3)$. The product would contain largely CH_4, if equilibrium were attained in all reactions. Therefore, the distribution of products should be of diagnostic value in interpreting the mechanism of chain growth.

About 1950, workers at the U.S. Bureau of Mines presented chain-growth schemes that predicted reasonably well carbon number and isomer distributions of aliphatic hydrocarbon from available data for the synthe-

---

[1] Current address: 111–58 Avenue S.W., Calgary, Alberta.

0-8412-0453-5/79/33-178-113$05.00/0
© 1979 American Chemical Society

sis on cobalt and iron 4, 5, 6, 7). Ref. 4, 5, and 6 summarize the literature to 1950. Weitkamp et al. (8) and Steizt and Barnes (9) reported
detailed analyses of products from a fluidized-bed synthesis on iron and
considered the chain-growth process (8).

One of the best of these schemes (10, 11), SCG (simple chain growth),
involved one-carbon addition to one end of the growing chain at the first
carbon with a probability a and to the second carbon with a probability
af, if addition had not already occurred on this carbon. The SCG scheme
assumed that the growth constants were independent of carbon number
and structure of the growing chain, a situation that is fortuitous rather
than expected. For a given carbon number, the ratios of branched to
normal hydrocarbons are f or 2f for monoethyl isomers and f^2 or $2f^2$ for
dimethyls; the factor one or two depends on whether the species can be
produced in one or two ways. This simple mechanism predicted carbon
number and isomer distributions moderately well and also seemed consistent with the tagged alcohol incorporation studies of Emmett, Kummer,
and Hall (12, 13, 14). SCG does not produce ethyl-substituted carbon
chains, which were subsequently found to be present in about the same
concentration as dimethyl species (15, 16).

Recently, Pichler and Schulz (16, 17, 18) have analyzed several
Fischer–Tropsch products by capillary gas chromatography, separating
virtually all of the species through C_9 and many of the isomers to C_{17} in
hydrogenated hydrocarbon fractions. These remarkable analytical data
permit a more detailed examination of the chain-growth processes.
Pichler and Schulz (17, 18) proposed a chain-growth scheme involving
surface species resembling carbonyls, an extension of an earlier postulate
of Wender (20, 21); however, tests have not been made on its ability to
predict the product distribution quantitatively. They demonstrated the
constancy of the major growth constant for products from cobalt catalysts
and obtained other interesting information (18, 19). Their objectives were
different from those of the present paper, and their work will not be
considered in detail here.

A recent paper presented a "Schulz–Flory polymerization" mechanism
for the synthesis (22). This scheme is a simplified form of SCG obtained
by setting the branching parameter f equal to zero.

In the present paper, more complicated chain-growth schemes, including a third constant, g, that is involved in producing ethyl-substituted
species, are proposed and tested. The two-constant scheme (SCG) is
easily developed, but in the three-constant schemes keeping track of all
the species generated becomes laborious. For this reason a method was
devised for performing this task with a CDC 6400 digital computer. In
this method the growing chain is defined by an array or vector of integers
with the number 1 denoting a single carbon; 2, a methyl-substituted

carbon; and 3, an ethyl-substituted carbon. This type of calculation may be useful for other catalytic processes, such as catalytic cracking, cf. the early work of Greensfelder, Voge, and Good (*23*).

Calculating the Distribution of Carbon-Chain Isomers

The carbon chain is represented by an array of integers or by a vector; for example, the carbon chain of 2-methyl-3-ethylpentane is represented by 1 1 3 2 1, where 3 denotes substitution of the ethyl group and 2, the methyl. A three-dimensional array ($N \times M \times I$) is used to designate all possible species. Actually each element of the vector representing a carbon chain consists of the array element $R(N, M, I)$, where N is the carbon number of the chain, M is an index of the isomers of carbon number N, and I is the position in the carbon chain. For each carbon number the index M is increased by one for each species formed, and M is used to examine the species in an orderly way in the next growth step. The value of I varies from 1 to 13, the first 12 representing the position along the carbon chain starting from the growing end, and the 13th, the probability of formation of the species.

Suppose, for example, 2-methyl-3-ethylpentane is the seventh ($M = 7$) C_8 species produced ($N = 8$) and its probability of formation is 0.05, then the array of numbers is: $R(8, 7, 1) = 1$; $R(8, 7, 2) = 2$; $R(8, 7, 3) = 3$; $R(8, 7, 4) = R(8, 7, 5) = 1$; $R(8, 7, 6) = R(8, 7, 7) = 0$; etc. to $I = 12$; and $R(8, 7, 13) = 0.05$.

The chain-growth rules are placed in the computer program. Table I shows nine different sets of rules; these rules will be described subsequently. Trial values of the growth constants are provided. For one-carbon additions (Schemes 1a–1d), the process starts with a C_3 species assigned a probability of one. New species ($N = 4$) are generated by adding carbon atoms one at a time according to the growth rules at only one end of the chain. For one- plus two-carbon additions, growth also occurs at only one end of the chain, but the calculation begins with a two-carbon species generating three- and four-carbon species in the initial growth steps.

In the present schemes, growth is permitted at only one end of the chain and the numbering for I (1–12) is from this end of the chain. For the simple one-carbon addition process (SCG),

$$111 \ (1.) \longrightarrow 1111 \ (a) \tag{1}$$

$$\searrow 121 \ (af) \tag{2}$$

where 1., a, and f are probability factors, and the resulting vectors are generated by the computer in the following order:
for Reaction 1:

$$R(4,1,I+1) = R(3,1,I) \quad \text{for } I = 1\text{–}11; R(4,1,1) = 1;$$
$$\text{and } R(4,1,13) = R(3,1,13)^* \, a;$$

for Reaction 2:

$$R(4,2,I) = R(3,1,I) \quad \text{for } I = 1\text{–}12,$$
$$R(4,2,2) = 2, \text{ and } R(4,2,13) = R(4,2,13)^* \, a \, {}^* \, f.$$

Here * means multiply.

For growth on the right end of the chain, the convention used in subsequent examples, the vectors for the species in the SCG scheme for Reactions 1 and 2 are:

N	M		I												
		13	12	11	10	9	8	7	6	5	4	3	2	1
3	1	1.	0	0	0	0	0	0	0	0	0	1	1	1
4	1	a	0	0	0	0	0	0	0	0	1	1	1	1
4	2	af	0	0	0	0	0	0	0	0	0	1	2	1

Starting with $N = 2$ or 3, the computer examines each species and applies the growth rules and then repeats the process for successively larger carbon numbers until the largest N desired is reached. After all growth steps have been made on all species of carbon number N, the carbon chain is rearranged so that the chain vector represents the longest carbon chain. For example, for $N = 6$ the following changes are made: 21111, 11112; 3111 and 1113, if present, are changed to 111111; and 1311 and 1131 become 11211. When the growing chain desorbs from the surface, certain distinct surface species become identical, for example, for $N = 7$ 121111 and 111121 become 2-methylhexane, 112111 and 111211 become 3-methylhexane, and 12211 and 11221 become 2,3-dimethylpentane; the probabilities for each of the species that become identical when the molecule leaves the surface are summed.

Finally, the predicted isomer distribution for carbon number N is calculated as $y_i = p_i / \sum_1^k p_j$, and a sum of squares of the differences between observed and predicted mole fractions $SQ_N = \sum_1^k (y_i' - y_i)^2$.

Table I.   Chain-Growth Rules[a]

Part 1—One-Carbon Addition

(1a) U—a→U 1   (all species)   1
       W 1 1—af→W 2 1   2
       X 1 1 1—ag→X $2^\dagger$ 1 1   3
       X $2^\dagger$ 1 1—a→X 3 1 1   4

(1b)   Same as (1a) except that (4) is
       X $2^\dagger$ 1 1—ag→X 3 1 1   4′

(1c)   Same as (1a) except that (3) and (4) are replaced by
       X 1 1 1—ag→X $2^\dagger 1^\dagger 1^\dagger$ 3″
       X $2^\dagger 1^\dagger 1^\dagger$—a→$X$ 3 1 1   (only)   4″

(1d) U—a→U 1   (all species)   1
       W 1 1 —af→W 2 1   2
       X 2 1 1—ag→X 3 1 1   3

After all growth steps on the species are completed, the following
transformations are made: $2 X = 1 1 X, 3 X = 1 1 1 X, 13 X =$
$1 1 2 X, X 2^\dagger 1 1 = X 2 1 1$, and $X 2^\dagger 1^\dagger 1^\dagger = X 2 1 1$.

Part 2—One- and Two-Carbon Additions

(2a) U—a→U 1   (all species)   1
       W 1—af→W 2   2
       W 1—a^2g→W 3   3
       W 1—a^2g→W 1 2   4

(2b)   An extension of (2a), same as (2a) except (4) is replaced by
       X—a^2g→X 2   (all species)   4′

(2c)   Same as (2a) except (3) is replaced by
       W 1—a^2fg→W 3   3′

(2d)   Same as (2a) except (4) is replaced by
       W 1—a^2f→W 1 2   4″

(2e)   An extension of (2c) ; all steps are shown.
       U—a→U 1   (all species)   1
       W 1—af→W 2   2
       W 1—a^2fg→W 3   3
       W 1—a^2g→W 1 2   4
       W 3—a^2fg→W 3 2   5
       W 2—a^2g→W 2 2   6

After all growth steps on the species have been completed, the
following transformations are made in this order: $W 2 = W 1 1$.
$W3 = W 1 1 1$, and $W 3 1 = W 2 1 1$.

[a] U, W, and X are parts of a carbon chain that may have any size or configuration.

Here y_i is the predicted mole fraction of isomer i of carbon number N, y_i' is the observed mole fraction of i, p_i is the predicted probability of isomer I as given in $I = 13$, and k is the number of isomers in carbon number N.

When the calculation has proceeded to the largest carbon number, N', an overall sum of squares is calculated, $\text{TSQ} = \sum_{n}^{N'} \text{SQ}_{N^1}$, where n is the smallest carbon number for which data are available.

Finally, a simple optimization program using a search procedure is used to adjust the values of the growth constants f and g to minimize TSQ, and the final predicted isomer distribution and "best" estimates of the growth constants are printed out.

Chain Growth Rules

Two types of chain growth patterns that produce ethyl-substituted carbon chains are presented in Table I, and each of them has two growth constants, f and g. Examples of how two of these schemes work are given in Tables II and III. Type 1 involves one-carbon additions to one end of the growing chain. For 1a, 1b, and 1c, addition is permitted at the first three carbons, and subsequent growth is permitted on the carbon atom added to the third carbon. In 1d, carbons are added to the first and second carbons and to a carbon substituted on the third carbon atom. As in SCG, addition to the second carbon is not permitted if one carbon already has been added to the second carbon. Growth rules of Type 1 are variants of the SCG (10, 11) and are reduced to the equations of SCG when $g = 0$.

Table II.   The Initial Part of Growth Scheme 1a[a]

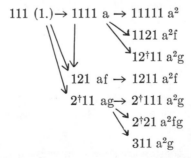

$$111 \ (1.) \rightarrow 1111 \ a \rightarrow 11111 \ a^2$$

Transformations after the growth steps: $2^{\dagger}11 = 1111$, $2^{\dagger}111 = 11111$, $2^{\dagger}21 = 1121$, and $311 = 11111$.

[a] Steps yielding species larger than C_5 are not shown.

Table III.   The Initial Part of Growth Schemes 2a and 2b[a]

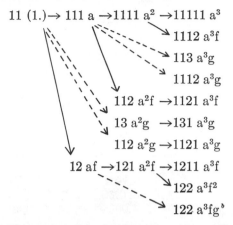

$$11\ (1.) \rightarrow 111\ a \rightarrow 1111\ a^2 \rightarrow 11111\ a^3$$

$$1112\ a^3f$$

$$113\ a^3g$$

$$1112\ a^3g$$

$$112\ a^2f \rightarrow 1121\ a^3f$$

$$13\ a^2g \rightarrow 131\ a^3g$$

$$112\ a^2g \rightarrow 1121\ a^3g$$

$$12\ af \rightarrow 121\ a^2f \rightarrow 1211\ a^3f$$

$$122\ a^3f^2$$

$$122\ a^3fg\ ^b$$

Transformations after the growth steps: $12 = 111, 112$ and $13 = 1111, 1112$ and $113 = 11111, 131 = 1121$, and $122 = 1211$.

[a] Steps yielding species larger than C_5 are not shown.
[b] This step is not permitted in 2a.

Growth rules of Type 2 are variants of the "alcohol dehydration" mechanism (*24, 25*) with one- and two-carbon units added at the one end of the growing chain. In 2a, 2c, and 2d, addition of two carbon units is not permitted when the growing end of the chain has the numbers 2 or 3, but this step is allowed in 2b and 2e.

Table II shows the initial part of Scheme 1a. Here $2^\dagger$ denotes a special carbon atom substituted on the third carbon, on which another carbon can be added. In the computer program, $2^\dagger$ is denoted by 9 and after all growth steps are completed for all species of this carbon number, the 9 is changed to 2 and all of the species are transformed as indicated so that the longest carbon chain is arranged along the vector. In Scheme 1c, $1^\dagger$ denotes a carbon atom to which another carbon cannot be added.

Table III shows the initial part of Scheme 2b which, for the species shown, is identical to 2a except that the reaction on the bottom of the table is not permitted in 2a. Transformations made after all growth steps on all species of a given carbon number are shown.

It is possible, but not necessary, to express the probabilities of forming each species in terms of constants f and g. The chain-growth calculation is made with $f = 0.1$ and $g = 0.01$ and with $f = 0.1$ and $g = 0.03$. The two sets of probabilities for all cases examined were sufficient to provide unique expressions for each component in terms of f and g. Evaluation of the probabilities for each species in terms of f and g saves computer time in the optimization step.

Table IV. Prediction of Isomer Distribution

Ratio of Branched-to-Normal Species

Car-bon No.	Isomer[a]	Expt.	One-Carbon Addition				
			1a	1b	1c	1d	SCG
4	2M	.0917	.0885	.0912	.0969	.1032	.1070
5	2M	.2330	.1935	.2355	.2126	.2064	.2140
6	2M	.2065	.1949	.2073	.1975	.2064	.2140
	3M	.1429	.1250	.1172	.1347	.1218	.1070
	23DM	.0113	.0094	.0107	.0095	.0107	.0113
7	2M	.1891	.1950	.2074	.1957	.2064	.2140
	3M	.2845	.2339	.2341	.2375	.2250	.2140
	23DM	.0306	.0226	.0242	.0229	.0233	.0226
	24DM	.0069	.0095	.0106	.0095	.0107	.0113
	3E	.0136	.0185	.0006	.0190	.0185	0
8	2M	.1941	.1949	.2074	.1956	.2064	.2140
	3M	.2294	.2338	.2340	.2376	.2250	.2140
	4M	.0970	.1088	.1168	.0987	.1033	.1070
	23DM	.0243	.0212	.0242	.0197	.0204	.0226
	24DM	.0186	.0227	.0243	.0228	.0232	.0226
	25DM	.0094	.0094	.0106	.0095	.0107	.0113
	34DM	.0131	.0137	.0136	.0136	.0127	.0113
	3E	.0280	.0372	.0011	.0400	.0372	0
	TM	.0037	.0010	.0010	.0010	.0012	.0012
	EM	.0037	.0036	.0001	.0038	.0038	0
9	2 + 4M	.3659	.4126	.4409	.3931	.4130	.4280
	3M	.2166	.2337	.2341	.2375	.2251	.2140
	23DM	.0218	.0213	.0242	.0193	.0214	.0226
	24DM	.0140	.0213	.0242	.0198	.0214	.0226
	25 + 35DM	.0377	.0359	.0379	.0365	.0359	.0339
	26DM	.0199	.0095	.0107	.0096	.0107	.0113
	34DM + 4E	.0417	.0442	.0278	.0427	.0418	
	3E	.0239	.0373	.0011	.0424	.0371	0
	TM	.0031	.0046	.0054	.0042	.0045	.0024

[a] M = methyl, DM = dimethyl, TM = trimethyl, E = ethyl, and EM = ethyl-methyl.

for Product from Entrained Iron

Ratio of Branched-to-Normal Species

		One- and Two-Carbon Addition		
2a	*2b*	*2c*	*2d*	*2e*
.0828	.0824	.0839	.0545	.0769
.2014	.2013	.2004	.1787	.2018
.1899	.1888	.2006	.1711	.1929
.1353	.1356	.1210	.1340	.1247
.0077	.0089	.0078	.0039	.0092
.1899	.1888	.2006	.1712	.1929
.2446	.2439	.2405	.2649	.2444
.0201	.0231	.0191	.0119	.0235
.0090	.0089	.0100	.0067	.0092
.0154	.0152	.0233	.0123	.0028
.1901	.1887	.2006	.1711	.1929
.2447	.2439	.2404	.2649	.2444
.1090	.1081	.1194	.1240	.1196
.0177	.0204	.0189	.0106	.0231
.0233	.0230	.0241	.0229	.0236
.0090	.0089	.0101	.0067	.0092
.0131	.0148	.0114	.0095	.0148
.0312	.0307	.0047	.0257	.0057
.0007	.0010	.0075	.0002	.0011
.0028	.0030	.0042	.0015	.0004
.4078	.4048	.4396	.4193	.4322
.2447	.2438	.2400	.2649	.2444
.0178	.0205	.0189	.0106	.0231
.0207	.0205	.0240	.0213	.0231
.0382	.0379	.0386	.0404	.0385
.0091	.0089	.0101	.0067	.0092
.0388	.0419	.0251	.0304	.0320
.0312	.0306	.0047	.0257	.0057
.0036	.0044	.0036	.0016	.0050

Application of Chain-Growth Schemes to Isomer Distributions

The detailed isomer distribution data of Pichler, Schulz, and Kühne (16, 17) for hydrogenated hydrocarbons from a fixed-bed synthesis on a precipitated cobalt catalyst at atmospheric pressure ($Co:ThO_2:kieselguhr = 100:18:100$) at 190°C and the entrained reactors of Sasol commercial plant in South Africa, using a reduced fused iron catalyst at 22 atm and about 320°C, were used for testing the nine chain-growth schemes in Table I in the range C_6–C_9. Before the analyses, the hydrocarbons were hydrogenated under conditions that should only saturate olefins. The

Table V. Predicted Isomer Distribution

Ratio of Branched-to-Normal Species

Car-bon No.	Isomer[a]	Expt.	One-Carbon Addition				
			1a	1b	1c	1d	SCG
4	2M	.0482	.0745	.0686	.0799	.0830	.0855
5	2M	.1390	.1567	.1576	.1664	.1660	.1709
6	2M	.1551	.1573	.1587	.1610	.1658	.1719
	3M	.0844	.0912	.0898	.0930	.0870	.0855
	23DM	.0012	.0061	.0061	.0064	.0069	.0073
7	2M	.1535	.1573	.1576	.1604	.1659	.1709
	3M	.1943	.1748	.1792	.1746	.1700	.1709
	23DM	.0041	.0137	.0141	.0139	.0142	.0146
	24DM	.0014	.0061	.0061	.0064	.0069	.0073
	3E	.0054	.0084	.0004	.0065	.0042	0
8	2M	.1487	.1573	.1577	.1604	.1658	.1709
	3M	.1811	.1748	.1793	.1746	.1700	.1709
	4M	.1001	.0835	.0894	.0804	.0829	.0855
	23DM	.0044	.0132	.0141	.0130	.0137	.0146
	24DM	.0074	.0138	.0141	.0139	.0140	.0146
	25DM	.0059	.0062	.0061	.0064	.0069	.0073
	3E	.0250	.0169	.0007	.0136	.0083	0
9	2 + 4M	.3428	.3245	.3365	.3213	.3316	.3418
	3M	.1785	.1749	.1793	.1746	.1699	.1709
	23DM	.0016	.0132	.0141	.0130	.0138	.0146
	24DM	.0095	.0132	.0141	.0130	.0138	.0146
	25 + 35DM	.0221	.0213	.0221	.0214	.0214	.0219
	26DM	.0032	.0064	.0062	.0065	.0068	.0073
	34DM + 4E	.0095	.0231	.0164	.0206	.0182	
	3E	.0126	.0168	.0008	.0143	.0082	0

[a] M = methyl, DM = dimethyl, and E = ethyl.

hydrocarbons from cobalt were all aliphatic, but those from the entrained iron synthesis contained, in some carbon number fractions, as much as 11% aromatics plus naphthenes. The composition of this product from entrained iron was calculated as the fraction of the total aliphatic hydrocarbon in a given carbon number for use in the calculations. Analyses are also available for products from a fixed-bed synthesis on precipitated iron (*16, 17*), but the present simple chain-growth schemes with growth constants independent of chain length and structure obviously would not represent these data because the fraction of the normal species does not decrease with increasing carbon number.

from Fixed-Bed Cobalt Catalyst

Ratio of Branched-to-Normal Species

One- and Two-Carbon Addition				
2a	*2b*	*2c*	*2d*	*2e*
.0717	.0721	.0652	.0454	.0648
.1602	.1601	.1547	.1397	.1543
.1555	.1557	.1548	.1410	.1532
.0951	.0946	.0923	.0978	.0925
.0057	.0060	.0046	.0022	.0059
.1544	.1557	.1548	.1410	.1532
.1794	.1787	.1836	.1979	.1825
.0129	.0139	.0111	.0070	.0140
.0060	.0061	.0060	.0046	.0058
.0061	.0057	.0014	.0012	.0014
.1554	.1557	.1549	.1410	.1532
.1794	.1788	.1835	.1979	.1824
.0843	.0841	.0913	.1000	.0901
.0121	.0132	.0110	.0070	.0137
.0139	.0139	.0142	.0141	.0140
.0060	.0060	.0060	.0045	.0058
.0121	.0113	.0028	.0025	.0028
.3240	.3240	.3375	.3411	.3332
.1794	.1788	.1834	.1979	.1826
.0122	.0132	.0110	.0070	.0138
.0132	.0132	.0142	.0141	.0138
.0219	.0221	.0226	.0239	.0223
.0060	.0060	.0059	.0045	.0059
.0202	.0207	.0145	.0112	.0178
.0122	.0133	.0029	.0024	.0027

The mole fraction of each isomer in a given carbon number was used in the calculation scheme, but the experimental and predicted data are reported in Tables IV, V, and VII as ratios of branched-to-normal species to keep the tables of reasonable size. These ratios for experimental results are reported to one more significant figure than original data to avoid rounding-off errors. Only the molecules reported in Ref. 16 and 17 were used in the comparison of predicted and experimental isomer fractions; this convention leads to no significant errors because the calculated fractions of molecules not reported were small. The optimization program was applied only to the C_6–C_9 isomers, and isomers in the C_4 and C_5 fractions were calculated using the growth constants obtained for C_6–C_9. For comparison, the isomer distributions also were calculated for SCG. Here an average value of f was obtained from the ratios of monomethyl to normal species from C_6–C_9.

Tables IV and V compare the predicted and experimental ratios of branched-to-normal molecules for products from iron and cobalt catalysts. Table VI presents the "best" values of the growth constants and the overall residual sum of squares, TSQ, for isomers in the range C_6–C_9. The chain-growth schemes generally predict the analytical data reasonably, often to within the experimental uncertainties of separating and identifying molecules in these complex mixtures. Formulating growth schemes that properly predict ethyl species was one of the objectives of this work. Schemes 1b, 2c, and 2e for both catalysts and 2d for cobalt predicted significantly lower amounts of ethyl species than observed, and SCG none. The best predicted results were obtained from Schemes 1a, 1c, 2a, and 2b. TSQ from SCG were calculated excluding ethyl species.

Table VI. Values of Growth Constants and TSQ for C_6–C_9 Isomers

Growth Scheme	Iron			Cobalt		
	f	g	TSQ[a]	f	g	TSQ[a]
1a	0.09018	0.01867	0.00290	0.07512	0.00848	0.00103
1b	.09336	.02344	.00451[b]	.06992	.01953	.00096[b]
1c	.09873	.01929	.00204	.08046	.00657	.00125
1d	.1032	.1800	.00320	.08291	.0500	.00144
2a	.09311	.01582	.00234	.07816	.00613	.00091
2b	.09260	.01550	.00226	.07849	.00567	.00097
2c	.09414	.02531	.00419[b]	.07129	.02000	.00070[b]
2d	.06205	.01326	.00471	.05004	.00125	.00107[b]
2e	.08635	.03330	.00369[b]	.0707	.0193	.00084[b]
SCG	.1070	—	.00546[c]	.0855	—	.00132[c]

[a] TSQ is the sum over all C_6 to C_9 species of the predicted mole fraction of each species minus the observed fraction, squared.
[b] Predicted values of ethyl species are too low.
[c] Ethyl species are not produced and are not included in calculating TSQ.

For SCG the values of TSQ would always be larger than the other growth schemes if the ethyl isomers were included in the calculations. The ratios for some species were predicted too large or too small by all schemes. For the product from cobalt, the predicted ratios for nearly all dimethyl species were two to four times the experimental ratios.

The new growth schemes also were applied to the older data (*11*); predictions from two schemes and SCG for iron and cobalt data are given in Table VII. The general pattern is predicted, but the deviations are large for iron and smaller for cobalt. For both data sets, Schemes 1a and 2a were better than SCG. Parameters for fluidized iron were similar to those for entrained iron. For cobalt the values of f were small and g was a sizable fraction of f.

Discussion

The present paper shows that a computer simulation can be made of the chain-growth process in the Fischer–Tropsch synthesis. The growing chain can be represented by an array of numbers, the characteristics of each species can be distinguished, the appropriate growth steps made to generate new species, and the results edited and summarized in a form that can be related to experimental data. Representing the carbon chain is simple because disubstitution on a single carbon, e.g., 2,2-dimethyl chains, 3,3-dimethyl, etc., is not permitted. However, each space in the *I* vector can accommodate a large number and 3.1, 4.1, and 5.2 could represent, respectively,

$$
-\overset{\displaystyle C}{\underset{\displaystyle C}{C}}-,\ -\overset{\displaystyle C}{\underset{\displaystyle C}{\overset{\displaystyle C}{C}}}-,\ \text{and}\ -\overset{\displaystyle C}{\underset{\displaystyle \underset{\displaystyle C}{C}}{\overset{\displaystyle C}{C}}}-\ .
$$

The present calculation methods through C_9 require a substantial amount of storage, approaching the capacity of a large computer even if only vectors for the last two or three carbon numbers are retained. The *I* vector for the Fischer–Tropsch synthesis could be stored as a number rather than an array of integers.

The one-carbon addition schemes are typical of processes in which the growing species may be envisioned to be a chemisorbed α olefin but with a finite probability that the double bond may move to β position. These schemes may be consistent with the recent mechanism of Pichler and Schulz (*17*). The one- and two-carbon-addition schemes are typical of the old "alcohol dehydration" mechanism, which incidentally has found no analogy in homogeneous organometallic reactions. Extension of the

Table VII.   Predicted Isomer Distribution for Older Data (*11*)

Fluidized Iron

Carbon No.	Isomer	Expt.	Calculated for Scheme		
			1a	*2a*	*SCG*
4	2M	.1186	.0944	.0974	.1061
5	2M	.2315	.2055	.2118	.2122
6	2M	.1421	.2073	.2010	.2122
	3M	.1206	.1316	.1398	.1061
	23DM	.0051	.0107	.0087	.0113
7	2M	.1985	.2073	.2011	.2122
	3M	.2894	.2469	.2549	.2122
	23DM	.0242	.0255	.0224	.0226
	24DM	.0045	.0107	.0101	.0113
8	2M 3M 4M	.5967	.5696	.5706	.5304
	23DM 24DM 25DM 34DM	.0426	.0752	.0700	.0675
	f		.09623	.09952	.1061
	g		.01909	.01517	—
	TSQ[a]		.00691	.00637	.00892

[a] For iron for C_4–C_8. For cobalt for C_6–C_8.

SCG scheme to permit two- as well as one-carbon addition should be tried. This growth pattern should be somewhat different from the schemes designated by 2.

The fact that constant growth parameters will predict the isomer distribution data reasonably is remarkable. It is not necessary that the kinetic constants governing chain growth are independent of chain length and structure but that certain ratios of these parameters are constant. The fraction of tertiary carbons has been reported to decrease with carbon number beyond C_{10} (*17*). The SCG scheme predicts a maximum and subsequent decrease, but the maxima occur at C_{12}–C_{14} for products considered in this chapter. For the cobalt product, all schemes predict yields of dimethyl species that are often too large by factors of two to four. The simple schemes with constant growth parameters as described here are unable to predict the isomer distribution sensibly for products from fixed-bed iron (*16*) and from fixed-bed nickel

Expressed as Ratio of Branched-to-Normal Species

Expt.	1a	2a	SCG
		Fixed-Bed Cobalt	
		Calculated for Scheme	
—			
.0526	.0575	.0667	.0733
.0636	.0579	.0572	.0733
.0525	.0517	.0539	.0367
—			
.0525	.0579	.0573	.0733
.0878	.0881	.0886	.0733
—			
—			
.0462	.0578	.0573	.0733
.0852	.0881	.0885	.0733
.0521	.0370	.0346	.0367
—			
	.02192	.02422	.03666
	.01508	.01031	—
	.000362	.000404	.00161

(*18*). Possibly the constant f of SCG scheme can be calculated for individual reaction steps as Schulz (*18, 19*) has done for the primary growth constant a.

Acknowledgment

The authors thank the National Research Council of Canada for supporting this work. We are grateful to T. W. Hoffman for introducing us to the intricacies of simulation of chemical plants.

Literature Cited

1. Anderson, R. B., Lee, C. B., Machiels, J. C., *Can. J. Chem. Eng.* (1976) **54**, 590.
2. Storch, H. H., Golumbic, N., Anderson, R. B., "The Fischer-Tropsch and Related Syntheses," Chap. 1, Wiley, New York, 1951.

3. Anderson, R. B., "Catalysis," P. H. Emmett, Ed., Vol. 4, Chap. 1, Reinhold, New York, 1956.
4. Storch, H. H., Golumbic, N., Anderson, R. B., "The Fischer-Tropsch and Related Syntheses," pp. 582–593, Wiley, New York, 1951.
5. Anderson, R. B., "Catalysis," P. H. Emmett, Ed., Vol. 4, pp. 345–367, Reinhold, New York, 1956.
6. Anderson, R. B., Friedel, R. A., Storch, H. H., *J. Chem. Phys.* (1951) **19**, 313.
7. Weller, S. W., Friedel, R. A., *J. Chem. Phys.* (1949) **17**, 801.
8. Weitkamp, A. W., et al., *Ind. Eng. Chem.* (1953) **45**, 343, 350, 359, 363.
9. Steizt, Alfred, Jr., Barnes, D. K., *Ind. Eng. Chem.* (1953) **45**, 353.
10. Storch, H. H., Golumbic, N., Anderson, R. B., "The Fischer-Tropsch and Related Syntheses," pp. 585–591, Wiley, New York, 1951.
11. Anderson, R. B., "Catalysis," P. H. Emmett, Ed., Vol. 4, pp. 353–359, Reinhold, New York, 1956.
12. Kummer, J. T., Emmett, P. H., *J. Am. Chem. Soc.* (1953) **75**, 5177.
13. Ibid. (1951) **73**, 564.
14. Hall, W. K., Kokes, R. J., Emmett, P. H., *J. Am. Chem. Soc.* (1960) **82**, 1027.
15. Blaustein, B. D., Wender, I., Anderson, R. B., *Nature* (1961) **189**, 224.
16. Pichler, H., Schulz, H., Kühne, D., *Brennst. Chem.* (1968) **49**, 1.
17. Pichler, H., Schulz, H., *Chem. Ing. Tech.* (1970) **42**, 1162.
18. Schulz, H., *Erdoel Kohle* (1977) **30**, 123.
19. Pichler, H., Schulz, H. H., Elstner, M., *Brennst. Chem.* (1967) **48**,·3.
20. Wender, I., Friedman, S., Steiner, W., Anderson, R. B., *Chem. & Ind.* (1958) **51**, 1964.
21. Sternberg, H. W., Wender, I., *Chem. Soc. (London)* **1959** Spec. Publ. No. 13, 55.
22. Henrici-Olivé, G., Olivé, S., *Angew. Chem. Int. Ed. Engl.* (1976) **15**, 136.
23. Greensfelder, B. S., Voge, H. H., Good, G. M., *Ind. Eng. Chem.* (1949) **41**, 2573.
24. Storch, H. H., Golumbic, N., Anderson, R. B., "The Fischer-Tropsch and Related Syntheses," pp. 592–593, Wiley, New York, 1951.
25. Anderson, R. B., "Catalysis," P. H. Emmett, Ed., Vol. 4, pp. 359–367, Reinhold, New York, 1956.

RECEIVED July 10, 1978.

10

Carburization Studies of Iron Fischer–Tropsch Catalysts

K. M. SANCIER, W. E. ISAKSON, and H. WISE

SRI International, Menlo Park, CA 94205

The carburization of two reduced iron catalysts in H_2/CO gas mixtures was studied in the temperature range 473–598 K by simultaneous measurements of mass gain and magnetic susceptibility. Under isothermal conditions the mass gain exhibited two separate regions of diffusion-rate-limited carburization. The ferromagnetic phases formed were found to depend on temperature and carburization time. At low temperatures, α-Fe was converted to $Fe_2C(hcp)$ and then to Fe_2C (Hägg). At the high temperatures, α-Fe was converted directly to Fe_2C (Hägg). Transitory formation of Fe_3C was observed. The buildup of free-surface carbon was noticeable after the iron was converted to Fe_2C (Hägg).

The properties of fully carburized iron Fischer–Tropsch catalysts have been the subject of several studies (*1, 2, 3, 4*). However the changes of the surface and bulk properties of the catalysts that accompany carburization in H_2/CO gas mixture over a range of temperatures have received less attention. During carburization of iron, the bulk reaction leads to the formation of different iron carbides which can undergo further solid-phase reactions. Both classes of reactions are temperature dependent (*1–7*), as summarized in Table 1. Our experiments were designed to examine the kinetics of carbon deposition and incorporation as well as the development of ferromagnetic phases in iron Fischer–Tropsch catalysts by simultaneous measurements of mass change and magnetic susceptibility during temperature-programmed and isothermal carburization.

0-8412-0453-5/79/33-178-129$05.00/0
© 1979 American Chemical Society

Table I.   **Principal Carburization Reactions of Iron and Phase Changes**

Carburization Reactions

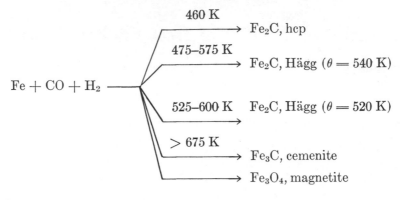

$$Fe + CO + H_2$$

460 K $\longrightarrow$ Fe_2C, hcp

475–575 K $\longrightarrow$ Fe_2C, Hägg ($\theta = 540$ K)

525–600 K $\longrightarrow$ Fe_2C, Hägg ($\theta = 520$ K)

> 675 K $\longrightarrow$ Fe_3C, cemenite

$\longrightarrow$ Fe_3O_4, magnetite

Phase Changes

$$Fe_2C, hcp \xrightarrow{616\ K} Fe_2C, Hägg$$

$$Fe_2C, Hägg\ (\theta = 540\ K) \xrightarrow{630\ K} Fe_2C, Hägg\ (\theta = 520\ K)$$

$$Fe_2C + Fe \xrightarrow{<\ 770\ K} Fe_3C$$

$$3Fe_2C \xrightarrow{770\ K} 2Fe_3C + C$$

$$Fe_3C \xrightarrow{873\ K} 3Fe + C$$

Experimental

Catalyst Samples. Samples of the iron catalysts B-2 and B-6 provided by ERDA–PERC were produced from a mixture of magnetite and the appropriate oxides, by calcining the mixture in an open iron crucible for 15 min in an induction furnace to 1740–1775 K (8). The chemical analyses of the catalysts after calcining are shown in Table II. The most significant difference between the catalysts is the high SiO_2 content in B-2. Most likely, the formation of iron silicate contributed to the high Fe^{2+}/Fe^{3+} in B-2 as compared to B-6.

Magnetic Susceptibility Apparatus. To study simultanously the rate of carbon buildup and the magnetic properties of the catalysts, we used a magnetic susceptibility apparatus equipped with an electronic microbalance (Cahn Model RS). We used a modified Faraday technique to measure magnetization, where the vertical magnetic-field gradient was provided by a set of electromagnetic gradient coils ($9, 10$) mounted on the flat pole faces of a 12-in. magnet. The assembly of the gradient coils, its power supply, and the hang-down tube with furnace for heating the sample in various gaseous atmospheres were purchased from George Associates (Berkeley, California). Two thermocouples were used; one in the gas stream just below the catalyst sample to measure the catalyst temperature and one near the furnace to control its temperature. The gradient coils were operated at a frequency 5 cps, and all measurements were normalized to a coil current of 1.00 A. The electrical signal from the microbalance was used to record mass changes and magnetization as a function of the magnetic field (0–8 kOe) or as a function of sample temperature (300–925 K), i.e., thermomagnetic analysis (TMA). For the TMA analysis, we used a lower field of 2.5 kOe to obtain more distinct Curie temperatures (11) and to decrease the effects of the magnet on the microbalance. For recording the magnetization, a lock-in amplifier was used to convert the 5-cps component in the microbalance signal to a dc voltage.

Catalyst samples (30 mg) were placed in a small quartz container suspended from the microbalance by a quartz fiber. The system was initially purged with helium (100 cm^3 min^{-1}) at room temperature. A separate stream of helium flowed continuously through the microbalance case. For reduction of the catalysts, hydrogen flow (100 cm^3 min^{-1}) was established, the temperature was programmed to rise at 18 K hr^{-1} from 450 to 725 K, the temperature was held at 725 K for three hours, during which time the catalyst weight became constant, and finally the sample was cooled in hydrogen. Changes in temperature and sample mass were recorded continuously during the reduction.

The reduced catalysts were carburized under either isothermal or temperature-programmed conditions at 1 atm. Unless otherwise stated, the carburizing gas was a mixture of H_2 and CO ($H_2/CO = 3$ by volume).

High-Pressure Tubular Reactor. For high-pressure carburization (1–10 atm), a tubular reactor was used. It was constructed of stainless steel (304 SS) with an internally mounted borosilicate glass frit to support the powdered catalyst and was designed so that the gas flowed into the reactor and through the catalyst bed before leaving the reactor. In a typical experiment, 0.090 g catalyst was loaded onto the glass frit of the high-pressure reactor. The sample was reduced in flowing H_2 (space velocity of 2×10^4 hr^{-1}) at 1 atm while the temperature was raised linearly from 450 to 725 K over a 16-hour period. The catalyst was subse-

Table II. Properties of Iron Oxide Catalysts B-2 and B-6

Property	Catalyst	
	B-2	B-6
Analysis (wt %) [a]		
Fe (total)	63.0	71.9
Fe^{+2}	59.0	29.5
Fe^{+3}	3.6	41.6
SiO$_2$	4.41	0.09
MgO	1.07	0.95
K$_2$O	0.36	0.27

[a] After calcination at 1740–1779 K (chemical analyses provided by ERDA–PERC).

quently cooled to 450 K and exposed to a flowing stream of $H_2/CO = 3/1$ (space velocity of 5×10^5 hr) at 1 or 10 atm while the temperature was programmed to rise from 450 to 575 K at a linear rate of 0.62 or 0.90 K min^{-1}. The catalyst was then cooled to room temperature in syngas and removed for analysis by TMA.

Gas Purification. The CO (Matheson "Ultra-pure" 99.8%) was purified of iron carbonyl by passing the sample through a copper tube packed with ⅛-in. Kaiser Al_2O_3 spheres and copper turnings and cooled in a bath of dry ice and acetone. Hydrogen and helium (Matheson, prepurified) were passed through traps in liquid nitrogen.

Mass Increase Resulting from Carburization. The increase of mass of hydrogen-reduced catalyst B-6 during a 72-hour carburization experiment under partial nonisothermal conditions is shown in Figure 1. During the first four hours, the temperature was programmed to rise from 425 to 598 K at a constant heating rate, after which it was held constant at 598 K.

An additional series of measurements were carried out under isothermal conditions at 1 atm (Figure 2). For these experiments, carburization of the hydrogen-reduced catalysts was limited to a period less than six hours. The data, analyzed in terms of the parabolic rate law, exhibit two distinct regions of carburization (Figure 3). The parabolic rate constants were calculated from the initial and final slopes (Table III). The parabolic rate constants yield the activation energies and preexponential factors summarized in Table IV.

Magnetization Resulting from Carburization. Thermomagnetic analysis (TMA) identified magnetite as the only ferromagnetic phase in the unreduced catalyst, and iron in the reduced samples. TMA measurements were made after isothermal carburization and after temperature-

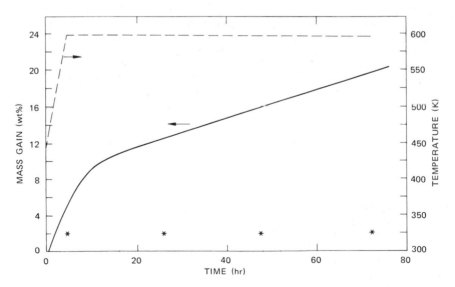

Figure 1. Mass gain during carburization of catalyst B-6 in H_2/CO (3/1) at 1 atm. Asterisk (*) indicates time at which thermomagnetic analysis was made.

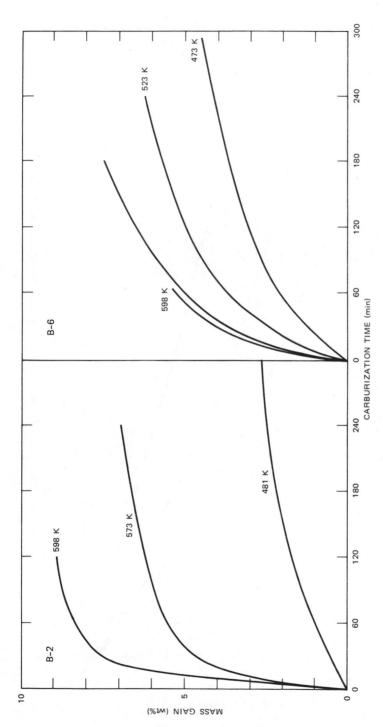

Figure 2. Mass gain during isothermal carburization of catalysts B-2 and B-6 in H_2/CO (3/1) at 1 atm

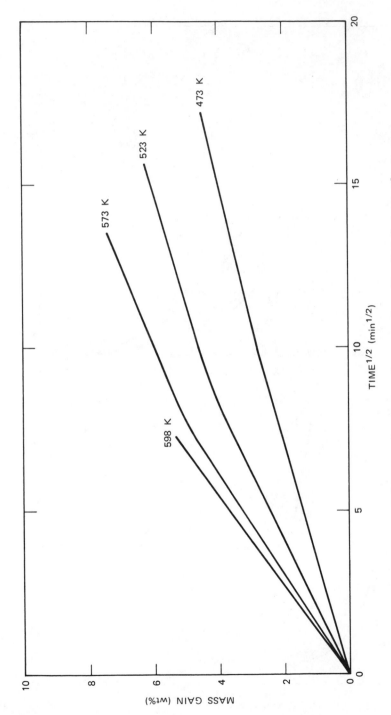

Figure 3.   Parabolic rate plot of mass increase of catalyst B-6 during carburization

Table III. Parabolic Rate Constants for Isothermal Carburization of Catalysts B-2 and B-6

Catalyst	Carburizing Temperature[a] (K)	Parabolic Rate Constant (wt % min$^{-1/2}$)	
		Initial Slope	Final Slope
B-2	481	0.15	—
	573	0.78	0.20 ± 0.02
	598	1.85	0.22
B-6	473	0.28	0.23
	523	0.48	0.29
	573	0.65	0.41
	597	0.72	—

[a] Carburization in H_2/CO (3/1) in 1 atm.

programmed carburization of catalysts B-2 and B-6. For these TMA analyses, we replaced the flowing syngas by helium and lowered the temperature from the reaction temperature to 350 K. From the Curie temperatures so evaluated, we assigned specific ferromagnetic phases in accordance with published data (Table V). The fraction of the total magnetization force attributable to a given phase was estimated by extrapolating the TMA curve of a given ferromagnetic phase to 300 K. The procedure for this estimation is illustrated in Figure 4 for TMA curves of carburized catalyst B-6 ($H_2/CO = 1.5$, 573 K, 4 hr). We obtained the upper TMA curve by raising the temperature of the sample from 350 to 950 K, and the lower curve by then lowering the temperature from 950 to 350 K.

The difference in the TMA curves can be accounted for by a reaction above 650 K between Fe_2C and Fe that resulted in formation of Fe_3C or decomposition of Fe_2C. The relative magnetization attributable to the various ferromagnetic phases can only be semiquantitatively estimated because of the unknown domain size, which at sufficiently small values gives rise to superparamagnetism rather than ferromagnetism (*12*).

Table IV. Activation Energy and Preexponential Factor for Mass Gain during Isothermal Carburization of Catalysts B-2 and B-6

Catalyst	Slope[a]	Activation Energy (kcal mol^{-1})	Preexponential Factor (wt % min$^{-1/2}$)
B-2	initial	10.8 ± 1.2	$1.3 (\pm 0.2) \times 10^4$
	final	—	—
B-6	initial	4.2 ± 0.3	26 ± 2
	final	3.2 ± 0.4	7 ± 1

[a] *See* Table III for conditions.

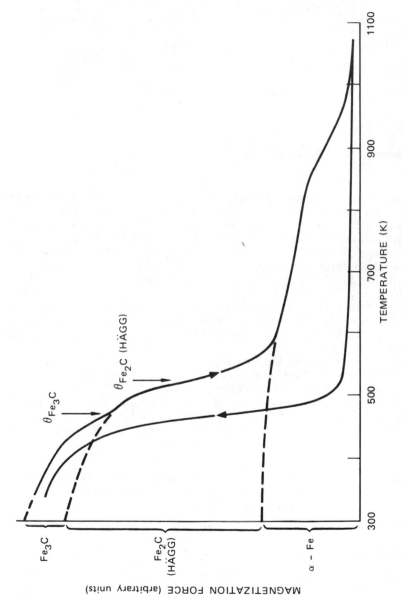

Figure 4. Typical thermomagnetic analysis (TMA) of carburized iron Fischer–Tropsch catalyst B-6

Table V. Curie Temperatures of Ferromagnetic Phases of Iron and Compounds

Phase	Curie Temperature (K)	References
$K_2O \cdot Fe_2O_3$	423	1
Fe_3C (cementite)	478–493	1,3
Fe_2C (Hägg)	520–540	1,2
Fe_2C (hcp) [a]	653	1
$Fe_{2.2}C$ (ϵ' phase)	723	2
Fe_3O_4 (magnetite)	838–868	1
α-Fe	1041	1

[a] Hexagonal close-packed.

The changes in the ferromagnetic phases during the 72-hour carburization of catalyst B-6 (Figure 5) were measured by TMA at the times indicated in Figure 1. As the α-Fe phase decreased, the Fe_2C (Hägg) phase increased, and the Fe_3C (cementite) phase exhibited a transient existence, reached a maximum at about five hours.

For the isothermal carburization of catalysts B-2 and B-6, the TMA results (Table VI) exhibit the following trends:

(a) For a given carburization temperature the total magnetization force was much greater for catalyst B-6 than for B-2.

(b) With increasing temperature the fractional contribution to the total magnetization by hexagonal, close-packed (hcp) Fe_2C and α-Fe decreased and that by Fe_2C (Hägg) increased. This latter effect is attributable to the higher stability of Hägg carbide over Fe_2C (hcp), as confirmed in a separate isothermal carburization of catalyst B-6 at 598 K. After a weight gain of 2.5 or 4.5 wt %, TMA revealed only Fe_2C (Hägg), not Fe_2C (hcp) or Fe_3C.

Table VI. Thermomagnetic Analysis of Catalysts B-2 and B-6 after Short–Term Isothermal Carburization

			Magnetic Properties		
	Carburization[a]	Total Magnetization	Ferromagnetic Phases, Fraction of M (%)		
Catalyst	Temperature (K)	Force, M (dyn g^{-1})	Fe_2C (Hägg)	Fe_2C (hcp)	α-Fe
B-2	481	100	0	50	50
	573	27	93	0	7
	598	56	89	0	11
B-6	473	340	0	59	41
	523	380	45	45	10
	573	495	87	0	13
	598	285	95	0	5

[a] Carburization in H_2/CO (3/1) at 1 atm.

Table VII. Effect of Heating Rate on Magnetic Properties of Catalysts

| | | *Total Magne-* | Ferromagnetic Phases Fe_2C (Hägg) | |
Catalyst	*Heating Rate*[a] *(K min⁻¹)*	*tization Force,* M *(dyn g⁻¹)*	*Curie Temperature (K)*	*Fraction of* M *(%)*
B-2	0.42	61	556	61
	0.62	196	570	63
	0.62 (T)[a]	180	598	70
B-6	0.37	72	549	58
	0.55	670	544	67
	1.00	830	520	94
	0.62 (T)[a]	350	617	87

[a] Carburization performed in magnetic susceptibility apparatus except for two runs in the tubular reactor (T).

(c) By carburizing the reduced catalysts, we did not produce the ferromagnetic phases $Fe_{2.2}C$ (ϵ' phase) and Fe_3O_4 (magnetite).

In the carburized catalysts, another ferromagnetic component, with a Curie temperature at about 423 K, was present. This component is probably associated with the $K_2O \cdot Fe_2O_3$ phase (1). Its magnetization was small, about 3–6% of the total, and its contribution was not included in the data in Table VI.

The effect of heating rate during carburization ($H_2/CO = 3/1$) on the magnetic properties of the catalysts were studied at 1 atm in the magnetic susceptibility apparatus and in the tubular reactor (Table VII). The mass gain of the samples carburized at 1 atm in the susceptibility

Table VIII. Effect of Pressure of $H_2/CO = 3/1$
Catalysts B-2 and B-6

| | | | Ferromagnetic Phases Fe_2C (Hägg) | |
Catalyst	*Total Gas Pressure*[a] *(atm)*	*Total Magnetic Force,* M *(dyn g⁻¹)*	*Curie Temperature (K)*	*Fraction of* M *(%)*
B-2	1	180	598	70
	10	175	617	69
B-6	1	400	617	87
	10	170	609	86

[a] Carburization in tubular reactor; temperature programmed at 0.62 K min⁻¹ from 465 to 575 K.

B–2 and B–6 during Carburization in H₂/CO = 3/1 at One Atmosphere

Magnetic Properties

	Ferromagnetic Phases		
Fe_2C *(hcp)*		*α-Fe*	
Curie Temperature (K)	*Fraction of M (%)*	*Curie Temperature (K)*	*Fraction of M (%)*
685	26	b	10
679	20	b	13
630	17	b	14
—	0	b	38
—	0	b	27
—	0	b	6
—	0	b	13

[b] Curie temperature of iron is above maximum temperature of TMA.

apparatus was 6.5 ± 0.2 wt %. To examine in more detail the effect of pressure on the formation of different bulk phases during carburization at a constant heating rate (0.62 K min⁻¹), we performed additional experiments in the tubular reactor (Table VIII). The TMA results indicate that:

(a) the total magnetization increased with heating rate;

(b) Fe_2C (Hägg) was the dominant ferromagnetic phase above 573 K;

(c) at comparable heating rates catalyst B-6 exhibited greater magnetization than did B-2;

(d) Fe_2C (hcp) was observed only in catalyst B-2; and

during Temperature–Programmed Carburization of on Magnetic Properties

Magnetic Properties

	Ferromagnetic Phases		
Fe_2C *(hcp)*		*α-Fe*	
Curie Temperature (K)	*Fraction of M (%)*	*Curie Temperature (K)*	*Fraction of M (%)*
630	16	b	14
662	17	b	14
—	0	b	13
—	0	b	14

[b] Curie temperature of iron above maximum temperature of TMA.

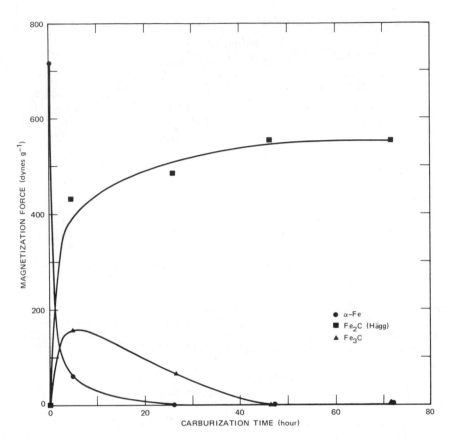

*Figure 5.   Ferromagnetic phases in catalyst B-6 during carburization. H₂/CO
(3/1) at 1 atm; for temperature see Figure 1.*

Table IX.   Effect of H₂/CO Ratio during Carburization

H_2/CO Ratio (vol %)	Carburization		
	Temperature (K)	Time (hr)	Mass Increase (%)
0^a	673^b	0.8	c
1.5	573	4.0	7.8
3.0	573	3.8	7.4
	673^b	6.0	44^e
4.0	573	4.5	5.3
	600	4.2	9.1

[a] He/CO = 1/1 (by volume).
[b] Carburized in tubular reactor.
[c] Mass change not measured; catalyst did not swell.

(e) for a given catalyst, the carburization pressure had a negligible effect on the relative amounts of the ferromagnetic phases produced.

The Curie temperature of a ferromagnetic phase which we tentatively assign to Fe_2C (Hägg) varied from 520 to 617 K (Table VII), although published data indicate a range of 520–549 K (Table V). The highest values of the Curie temperature for Fe_2C (Hägg) were obtained by carburizing in the tubular reactor. Also, in some instances the Curie temperature measured from Fe_2C (hcp) was higher than the reported value of 653 K. Therefore, the Curie temperature appears to be affected by the catalyst composition and by the heating rate during carburization, which in turn determine the actual catalyst temperature.

Effect of Carburizing Gas Composition. After reduction in hydrogen, catalyst B-6 was carburized at several temperatures in a gas containing various H_2/CO ratios, and the bulk phases were determined by x-ray diffraction and TMA. Carburization at 673 K was carried out in the tubular reactor. The bulk-phase composition depended both on gas composition and temperature (Table IX). Fe_3C (cementite) predominated when catalyst B-6 was exposed to CO, whereas when it was exposed to H_2/CO mixtures, Fe_2C (Hägg) was the major phase. Larger mass increases were observed at higher temperatures in gas mixtures of H_2 and CO.

Relationship between Magnetization and Mass Increase during Carburization. The relationship between total magnetization and mass increase was investigated during isothermal (598 K) carburization ($H_2/CO = 3/1$) of reduced catalyst B-2. The results are summarized in Figure 6 in which the mass increase identified by the carbon/iron atom ratio is indicated on the abscissa, and the percent of the initial magnetization force attributable to α-Fe at 598 K is plotted on the ordinate. Since the reaction temperature is above the Curie temperatures of Fe_2C (Hägg) and Fe_3C, and Fe_2C (hcp) is unstable at this temperature, the magnetization measured was attributable almost exclusively to α-Fe. The experi-

at One Atmosphere on Magnetization of Catalyst B-6

Total Magnetization Force, M $(dyn\ g^{-1})$	Magnetic Properties		
	Ferromagnetic Phases, Fraction of M (%)		
	Fe_3C	Fe_2C (Hägg)	α-Fe
—	100[d]	0	0
590	10	63	27
400	0	87	13
—	10	90[f]	0
515	13	73	14
480	36	54	10

[d] X-ray analysis indicates Fe_3C (cementite) only.
[e] Catalyst swelled in volume.
[f] X-ray analysis indicates Fe_2C (Hägg) and carbon as major phases.

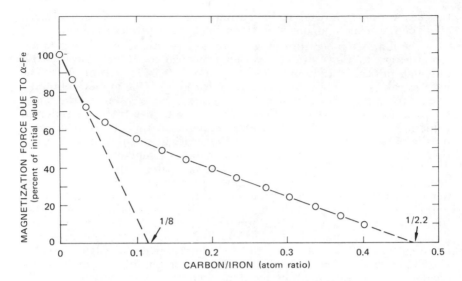

Figure 6.   Magnetization force attributable to α-Fe as a function of carbon/iron ratio during carburization of reduced catalyst B-2 at 598 K

mental points appear to follow two lines with different slopes (Figure 6) whose intercepts with the abscissa correspond to a carbon/iron ratio (C/Fe = 1/8 and C/Fe = 1/2.2).

Discussion

Carburization Kinetics. The trends in the mass increase during carburization of reduced B-2 and B-6 catalysts (Figure 1) are qualitatively similar to those reported (*1, 6, 13*). The rapid initial mass gain, presumably attributable to carbide formation, is followed by a lower, constant rate of increase that is attributed to free carbon (*6*). The formation of the free carbon, which has been associated with catalyst swelling, occurred only on exposure to H_2/CO mixtures but not CO alone and when the Fe_2C (Hägg) phase predominated.

Information on the mechanism of carburization can be obtained from the mass increase observed under conditions of isothermal carburization. The parabolic rates observed (Figure 3) suggest that carburization proceeded in two distinct diffusion-limited regimes. The observation of two diffusion-controlled regimes is supported by the combined study of mass increase and magnetization during carburization (Figure 6). The intercept corresponding to the Fe/C ratio of eight suggests a ferromagnetic phase of iron carbide with low carbon content. The intercept of Fe/C = 2.2 is close to the value expected for Fe_2C (Hägg), the predominant ferromagnetic phase in this sample after carburization.

To account for the carburization of iron we propose the following four-step mechanism:

(1) dissociative chemisorption of CO;

(2) during the initial parabolic regime, bulk diffusion of carbon results in a ferromagnetic iron–carbide phase which is low in carbon;

(3) during the latter parabolic regime, further bulk incorporation of carbon results in the formation of Fe_2C (Hägg); and

(4) buildup of free-surface carbon.

The rates of these various steps will depend on carburization conditions and catalyst composition. That the activation energy for mass increase during carburization is higher for catalyst B-2 than for B-6 (Table IV) may be associated with the greater amount of SiO_2 in B-2.

Carbon buildup at 598 K became clearly evident when the bulk composition attained Fe_2C, mostly Hägg carbide. Similar behavior was observed by Storch et al. (1), who reported that the amount of carbidic carbon became constant after carburizing in 0.1 atm CO at 598 K for three hours and by Hofer et al. (13) during carburization in CO at lower temperatures. Turkdogen and Vinter (14) studied carburization of iron in CO and H_2/CO mixtures, and they reported that the rate of carbon deposition increased with temperature in the range 673–1073 K and that the carbon deposition ceased when the iron was converted to cementite.

Ferromagnetic Phases. Our TMA results indicate that the ferromagnetic phases produced by carburizing the iron catalysts depend primarily on carburization temperature (Table I), and, to some extent, on the H_2/CO ratio. When carburizing in $H_2/CO = 3/1$ with increasing temperature, we observed the following:

(a) α-Fe decreased;

(b) Fe_2C (hcp) increased, but above 573 K it was unstable with respect to Fe_2C (Hägg), which was the predominant phase at carburization temperatures up to 598 K; and

(c) Fe_3C (cementite) was produced under special conditions.

At the highest temperature (673 K), Fe_3C formation was favored by carburizing in CO, but Fe_2C (Hägg) was favored when both H_2 and CO were present (Table IX). Also, Fe_3C was formed and then disappeared in the 72-hour carburization experiment (Figure 5). The transitory formation of Fe_3C has been reported previously (15).

We did not observe two ferromagnetic phases, $Fe_{2.2}C$ (ϵ' phase) and Fe_3O_4, as previously observed by TMA in carburized iron (1, 2). This difference in the studies is probably attributable to differences in catalyst compositions and reaction conditions. In a few but not all instances of Ref. 2, the assignment of Fe_3O_4 of an inflection at about 825 K in the TMA curve appears to be incorrect because the inflection disappeared upon cooling. This disappearance is a strong indication of a thermal reaction between Fe and Fe_3C.

The large range of Curie temperatures exhibited by our carburized B-6 catalysts (520–617 K) and the total magnetization appear to be associated with the thermal history of the sample, especially the heating rate during carburization. The data indicate that low Curie temperature and high magnetization accompany high heating rate. This effect may be associated with the higher temperatures attained by the catalyst at large heating rates. Under these conditions, the exothermic heat of the carburization reaction cannot be as readily dissipated from the sample to the environment. Similarly, the lower Curie temperatures of the Fe_2C (Hägg) (520–570 K) produced by carburizing in the magnetic susceptibility apparatus relative to the tubular reactor (598–617 K) may be the result of faster heat transfer from the catalyst in the tubular reactor (Table II).

The variable Curie temperatures observed in our study can be ascribed to new ferromagnetic phases or to a single phase with variable degrees of crystalline imperfection. At high heating rates during carburization, the resulting high catalyst temperature favored annealing of crystallite imperfections and an increase in the size of ferromagnetic domains, which in turn resulted in greater magnetization. Three ferromagnetic phases in carburized iron have been reported to have Curie temperatures in the range 520–653 K. Two modifications of Hägg Fe_2C carbides were characterized by x-ray diffraction with Curie temperatures from 513 to 523 K and 533 to 543 K (5). The third ferromagnetic phase is the well-characterized hexagonal Fe_2C with a Curie temperature of 653 K. To account for the existence of two modifications of Hägg Fe_2C, Cohn et al. (5) suggested differences in crystalline imperfections resulting from small crystallites or lattice strain.

The minor ferromagnetic phase that was evident in some samples from the Curie temperature of 423 K was assumed to be attributable to potassium ferrite, $K_2O \cdot Fe_2O_3$ ($\theta = 423$ K) (1). The amount of this phase indicates that the K_2O that was added as a promoter is mostly tied up with an iron oxide.

Acknowledgment

Support of this research by the Department of Energy (Contract No. E(36-2)-0060) is gratefully acknowledged.

Literature Cited

1. Storch, H. H., Golumbic, N., Anderson, R. B., "The Fischer-Tropsch and Related Syntheses," Wiley, New York, 1951.
2. Loktev, S. M., Makarenkova, L. I., Slivinskii, E. V., Entin, S. D., *Kinet. Catal.* (1972) **13**, 933.
3. Maksimov, Y. V., Suzdatev, I. P., Arents, R. A., Loktev, S. M., *Kinet. Catal.* (1974) **15**, 1144.

4. Amelse, J. A., Butt, J. B., Schwartz, L. H., *J. Phys. Chem.* (1978) **82**, 558.
5. Cohn, E. M., Bean ,E. H., Mentser, M., Hofer, L. J. E., Pontello, A., Peebles, W. C., Jacks, K. H., *J. Appl. Chem.* (1955) **5**, 418.
6. Pichler, H., Merkel, H., "Chemical and Thermomagnetic Studies on Iron Catalysts for Synthesis of Hydrocarbon," U.S. Bureau of Mines, 1949, Technical Paper 718.
7. Stoner, E. C., "Magnetism and Matter," pp. 280–434, Methuen, London, 1934.
8. Mauser, J. E., private communication, Bureau of Mines, Albany Metallurgy Research Center, Albany, NY.
9. Lewis, R. T., *Rev. Sci. Instrum.* (1971) **42**, 31.
10. Lewis, R. T., *J. Vac. Sci. Technol.* (1974) **11**, 404.
11. Kussman ,A., Schulze, A., *Phys. Z.* (1937) **38**, 42.
12. Selwood, P. W., "Adsorption and Collective Paramagnetism," Chap. 3, p. 35–50, Academic, New York, 1962.
13. Hofer, L. J. E., Cohn, A. M., Peebles, W. C., *J. Am. Chem. Soc.* (1949) **71**, 189.
14. Turkdogen, E. T., Vinters, J. V., *Metall. Trans.* (1974) **5**, 11.
15. Pichler, H., Schulz, H., *Chem. Ing. Tech.* (1972) **42**, 1162.
16. Anderson, R. B., Hofer, L. J. E., Cohn, E. M., Seligman, B., *J. Am. Chem. Soc.* (1951) **73**, 944.

RECEIVED June 22, 1978.

Rhodium-Based Catalysts for the Conversion of Synthesis Gas to Two-Carbon Chemicals

P. C. ELLGEN, W. J. BARTLEY, M. M. BHASIN, and T. P. WILSON[1]

Union Carbide Corp., Research and Development Department, Chemicals and Plastics, P.O. Box 8361, South Charleston, WV 25303

Methane, acetic acid, acetaldehyde, and ethanol constitute approximately 90 carbon atom percent of the primary products from the hydrogenation of CO over Rh/SiO_2 and $Rh–Mn/SiO_2$ catalysts at 250°–300°C and 30–200 atm pressure in a back-mixed reactor with $H_2/CO = 1$. The rate of reaction and the ratio, CH_4/C_2 chemicals, vary with $(P_{H_2}/P_{CO})^{0.5}$. The addition of 1% Mn raises the synthesis rate of a 2.5% Rh/SiO_2 catalyst about tenfold. The kinetics and the product distribution are consistent with a mechanism in which CO is adsorbed both associatively and dissociatively. The surface carbon produced by the dissociative CO chemisorption is hydrogenated through a $Rh–CH_3$ intermediate, and CO insertion in that intermediate results in formation of surface acetyl groups.

This symposium on advances in Fischer–Tropsch chemistry is testimony to the resurgence of interest in the field resulting from 1973 price increases and the prospect of reduced availability of petroleum-derived raw materials. The technological bases for current work in synthesis gas conversion are largely found in pioneering experiments done between 1930 and 1960 (*1, 2*). The more recently developed tools and concepts of heterogeneous catalysis have been applied to the problems of modifying the catalysts used earlier, in efforts to devise more economical synthesis gas-based processes. The problems with the older catalysts can be summarized in the statement that they generally produced broad mixtures of products, principally hydrocarbons. It would be preferable

[1] Author to whom correspondence should be addressed.

0-8412-0453-5/79/33-178-147$05.00/0
© 1979 American Chemical Society

to retain in the product the oxygen originally associated with the carbon monoxide converted, because both the weight and the value of such a product would be greater than that of the corresponding hydrocarbon. Furthermore, production of a single chemical, as in methanol production, confers substantial economic benefits.

Relatively little information is available on the characteristics of group VIII metals other than ruthenium and the members of the first row, although all have been investigated to some extent (3, 4, 5). A screening study of the lesser known members of that group (6, 7) gave evidence that rhodium exhibited a unique ability to produce two-carbon chemicals—acetic acid, acetaldehyde, and ethanol. Studies of supported rhodium and rhodium–iron catalysts showed that iron additions increased the production of methanol and ethanol at the expense of acetic acid and acetaldehyde (8). Results of further investigations of supported rhodium catalysts will be discussed here. The promoting effects of added manganese salts (9) will be shown. The results of kinetic studies will be described for synthesis gas conversion over rhodium and rhodium–manganese catalysts. These results will be compared with those available in the literature for related systems, and some mechanistic speculations will be presented.

Experimental

Details of the experimental techniques have been reported elsewhere (6, 7, 8, 9). Catalysts were prepared by conventional incipient wetness techniques using $RhCl_3 \cdot xH_2O$ and $Mn(NO_3)_2$ in aqueous solution. All catalysts described here were prepared on Davison Grade D59 silica gel, dried in air, and reduced in flowing hydrogen, using a programmed increase in temperature ending with one hour at 500°C.

Carbon monoxide chemisorption was measured at 25°C on all catalysts tested. The results were used in calculations of turnover numbers, assuming only rhodium metal atoms chemisorbed CO. While this assumption is naive and probably incorrect (10), results correlated moderately well with rhodium particle size determinations by transmission electron microscopy.

Catalysts were tested in a flow system at conversions generally below 10%. Reactors were back-mixed, Berry-type units, plated with gold or silver to minimize carbonyl formation and consequent catalyst contamination. Liquid and gaseous products were analyzed by VPC, using procedures described elsewhere (6, 7, 8).

Results

Data presented in Figures 1 and 2 show the activities (rates of CO conversion) obtained with 2.5% Rh/SiO_2 catalysts and with similar catalysts to which varying quantities of manganese had been added.

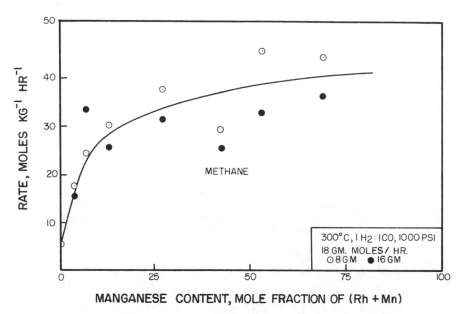

Figure 1. Effect of manganese on rates to methane. 300°C, 1 H$_2$:1 CO, 1000 psi, 18 gm mol/hr. (⊙) 8 gm, (●) 16 gm catalyst.

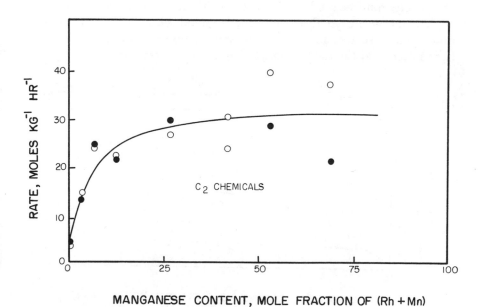

Figure 2. Effect of manganese on rates to C$_2$ chemicals

Table I. Effects of Operating

Power Law Parameters

Catalyst	Molar Rate of Product Formation			
	$\Delta E_{act.}$	X	Y	Mult. r^2
5% Rh, D59SiO$_2$	26 ± 0.7	0.87 ± 0.08	-0.40 ± 0.08	0.99
1.8% Rh–0.8% Mn, D59 SiO$_2$	24 ± 2	0.58 ± 0.12	-0.48 ± 0.12	0.93
2.5% Rh–1.0% Mn, washed D59 SiO$_2$	24 ± 2.5	0.64 ± 0.1	-0.33 ± 0.1	0.92

[a] Range of variables: $T = 285°–315°C$, $P_{H_2} = 28–62$ atm, $P_{CO} = 28–62$ atm, space velocity = 15,000–45,000 hr^{-1}. The error figures shown are the calculated standard errors of the estimates. Mult. r^2 = the fraction of the total variance in the data accounted for by the equation tested. *See* text for meaning of symbols.

The pronounced increase in rate of production of CH$_4$ and C$_2$ chemicals (acetic acid, acetaldehyde, and ethanol) as a result of the addition of relatively small quantities of manganese is evident. The selectivity of the reaction was not particularly sensitive to additions of manganese, as shown in Figure 3. There the total carbon atoms in products other than CO$_2$ are divided up into bands to indicate the percentage represented by the several products. There is some question as to whether or not a completely iron-free rhodium or rhodium–manganese catalyst would make any ethanol at all. As discussed in more detail elsewhere (8), traces of iron could have remained in the washed supports used or could have been introduced through carbonyls formed in the reactor.

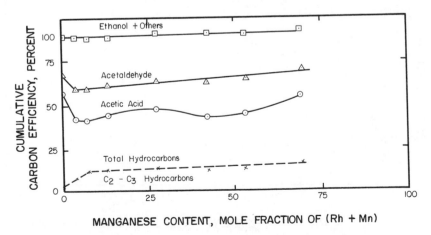

Figure 3.   Selectivities of rhodium-manganese/silica gel catalysts

Variables on Catalyst Performance[a]

	Power Law Parameters		
	CH_4/C_2 Chemicals (Mole Ratio)		
$\Delta E_{act.}'$	X'	Y'	Mult. r^2
—	—	—	—
3.8 ± 3	0.23 ± 0.1	-0.60 ± 0.1	0.80
10.1 ± 2.5	0.66 ± 0.1	-0.41 ± 0.1	0.84

Molar rates of product formation were determined as functions of temperature, P_{H_2}, P_{CO}, and space velocity in factorially designed sets of experiments on three different rhodium-based catalysts. Results are reported in Table I. A selectivity characteristic, the mole ratio of methane to the sum of the C_2 chemicals produced, also is reported there for two of the three catalysts. Both the rate and the selectivity characteristics are assumed to be representable by a power law equation of the type:

$$\text{variable} = K \exp(-\Delta E_a/RT) P_{H_2}{}^X P_{CO}{}^Y$$

Values of $\Delta E_a(\text{kcal/g mol})$, X and Y are given in the table, along with a number designated Mult. r^2 which is a statistical measure of the precision with which the equation represents the experimental data. Rather than reporting values of K, turnover numbers have been calculated for a temperature of 300°C and 3 H_2:1 CO synthesis gas (note the difference in pressures). These are reported in Table II, where they are compared with related data from the literature (11, 12).

Discussion

The overall reaction of synthesis gas over Rh and Rh–Mn catalysts, under the conditions of the experiments reported here, is represented in general terms by:

$$CO + H_2 \rightarrow CH_4 + CH_3COOH + CH_3CHO + H_2O$$

It will be argued that this reaction proceeds through the steps shown in Figure 4.

	Table II. Synthesis
	Turnover Numbers
Work by:	V(11)
Conditions:	1 atm, 2 H_2:1 CO
	Rh/Al_2O_3
% Carbon efficiency to:	
CH_4	90
C_2 + HC	10
C_2 chemicals	—
Turnover no., sec.$^{-1}$ (300°)	0.034

^{a} Calculated—3 H_2:1 CO.

The effect of added manganese is to raise the rate of product formation by a rhodium catalyst by approximately one order of magnitude. Despite this, neither the temperature nor the pressure dependence of the reaction rate is changed significantly by the addition. Furthermore, the temperature and pressure dependences of the rate of product formation in these experiments are quite close to those determined by Vannice (11) for methane formation over a Rh/Al_2O_3 catalyst at atmospheric pressure. Vannice reports that his methanation results can be represented by:

$$rate = A \exp(-24{,}000/RT) P_{H_2}^{1.04} P_{CO}^{-0.20}$$

Our own best data on molar rate of formation of all products are represented by:

$$rate = B \exp(-24{,}000/RT) P_{H_2}^{0.64} P_{CO}^{-0.33}$$

The turnover numbers shown in Table II for our Rh/SiO_2 and for Vannice's Rh/Al_2O_3 experiments are of the same magnitude. If the pressure differences are corrected on the assumption that the average dependence on total pressure over the pressure range of interest is proportional to $P^{0.5}$ (Vannice, $P^{0.8}$; EBBW, $P^{0.3}$), the turnover numbers measured in the experiments reported here are lower than those reported by Vannice by a factor of 5.5. Given the differences in the catalysts, the numbers are reasonably close. On that basis, it will be assumed in the subsequent discussion that the two investigations studied the kinetics of the same rate-determining process under different conditions. Clearly there is a difference in the amount of C_2 chemicals produced, though.

Sexton and Somorjai (12) have reported results of Auger and other studies on a polycrystalline rhodium surface exposed to CO, CO_2, and H_2. They found that after reaction of H_2 and CO with the specimen at

Gas Conversion

Turnover Numbers		
S and S (12)	*This Work*	
1 atm, 3 H_2:1 CO	70 atm, 1 H_2:1 CO	
Rh	Rh/SiO$_2$	Rh–Mn/SiO$_2$
90	53	30
10	4	12
—	43	58
0.13	0.030 (0.052) [a]	0.32 (0.55) [a]

one atmosphere and at 300°C, the surface became coated with a mono-layer of a chemically reactive carbonaceous deposit. The carbon could be removed as methane by treatment with hydrogen at the same tem-perature. This behavior is essentially the same as that reported for nickel catalysts by Wentrcek, Wood, and Wise (13) and by Araki and Ponec (14). The latter authors pointed out that the kinetics reported by Van-nice (11) for methanation reactions over supported group VIII metal catalysts could be interpreted as well on the basis of a mechanism

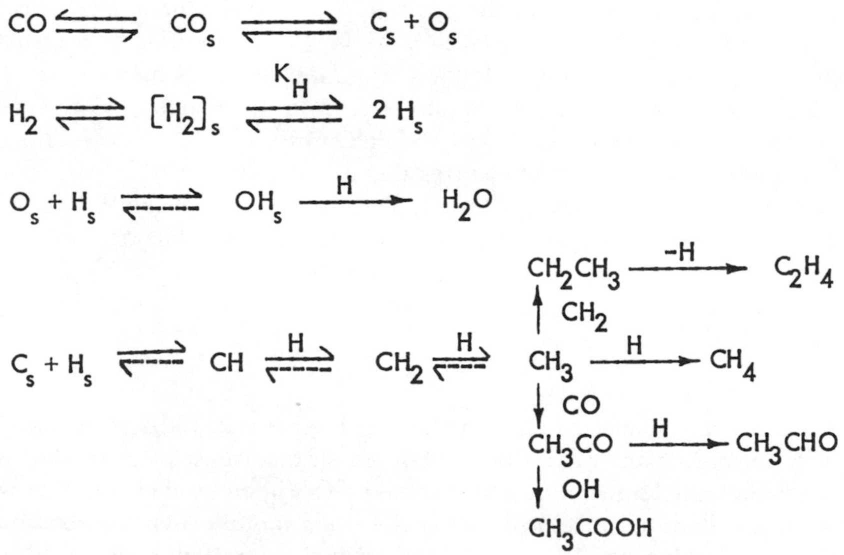

Figure 4.   *Proposed mechanism of synthesis gas conversion over rhodium catalysts*

involving stepwise addition of hydrogen atoms to carbonaceous intermediates on the nickel surface as they could be by the hydroxy–carbene mechanism used by Vannice (*15*) and by Vannice and Ollis (*16*) in analyzing the results.

Araki and Ponec (*14*) concluded that the rate of methane formation from carbon deposited on the surface by the dissociative chemisorption of CO could be represented by:

$$r = k\theta_C\theta_H{}^m \tag{1}$$

Here m is the number of hydrogen atoms in the product of the rate-determining step (RDS). It is assumed that the reactions preceding the RDS are at equilibrium, so the concentrations of the intermediates involved in those reactions are related by equilibrium constants to θ_C. For example, if CH reacts with H in the RDS,

$$\theta_{CH} = K_1\theta_C\theta_H \tag{2}$$

and

$$r = k_2\theta_{CH}\theta_H = k_2K_1\theta_C\theta_H{}^2 \tag{3}$$

A reasonable idealization of the kinetic models given in Table I is that in the high pressure limit both the synthesis rate and the selectivity ratio are proportional to $P_{H_2}{}^{1/2} \cdot P_{CO}{}^{-1/2}$. Although several assumptions are required, this result can be rationalized in terms of a mechanism similar to those proposed previously (*13, 14, 15, 16*) for related systems. The first of these assumptions follows from Sexton and Somorjai's results (*12*), which suggest that a constant coverage of the rhodium surface by carbon would be achieved under the high pressure synthesis conditions. A chemical model which rationalizes this assumption is that a particular class of surface sites always is essentially occupied completely by carbon atoms because dissociative chemisorption of CO is a kinetically facile process at those sites. Under this assumption, Equation 1 reduces to Equation 4.

$$r = k_m\theta_H{}^m \tag{4}$$

A second assumption is that CO and dihydrogen absorb competitively in molecular form on the remaining surface sites. Because Sexton and Somorjai's temperature-programmed desorption studies (*12*) show that molecularly chemisorbed CO is desorbed rapidly from the rhodium surface only above 250°C at 10^{-8} Torr, virtual saturation of these associatively adsorbing sites by carbon monoxide can be assumed to occur at 300°C and 50 atm. The fact that molecular hydrogen does not compete

effectively with CO for chemisorption on rhodium is clearly the fundamental factor behind the observed inhibition of synthesis rate by CO.

Now, assuming that adsorbed hydrogen atoms are formed via the
equilibrium processes shown in Figure 4,

$$\theta_H = K_H^{1/2}\theta_{H_2}^{1/2}, \tag{5}$$

and substitution of the appropriate Langmuir–Hinshelwood expression
for θ_{H_2} gives:

$$\theta_H = \frac{aK_H^{1/2}P_{H_2}^{1/2}}{(1 + bP_{CO})^{1/2}} \tag{6}$$

which reduces at high CO coverage to

$$\theta_H \propto P_{H_2}^{1/2} P_{CO}^{-1/2} \tag{7}$$

Substitution of Equation 7 in Equation 4, with $m = 1$, gives a result in
satisfactory agreement with the experimental high pressure results. Thus,
for both Rh/SiO_2 and $Rh–Mn/SiO_2$ catalysts, this analysis implies that
the first addition of H_s to C_s is the rate-determining step. It appears
that manganese influences the rate of the rhodium catalyst's rate-determining step, but not sufficiently to shift the synthesis rate limitation to
another elementary process.

Vannice's data (*11*) on rate of methane formation over a Rh/Al_2O_3
catalyst can be interpreted in terms of Equation 3, which is equivalent
to Equation 4 with $m = 2$, as well as in terms of the mechanisms suggested by Vannice (*15*) and Vannice and Ollis (*16*). Substitution of
Equation 6 in Equation 4 yields:

$$r = \frac{k_2K_HP_{H_2}}{1 + bP_{CO}} \tag{8}$$

Vannice's CO partial pressures were 0.08–0.24 atm, and his temperatures were 240°–280°C. Equation 8 gives the observed values for X
(1.04) and Y (−0.20) if $b = 2$ atm^{-1}. That value of b cannot be directly
transferred to the data presented here because of the difference in average
reaction temperatures in the two sets of experiments. It does, however,
make plausible the implication of the power law coefficients reported in
Table I that the surface is very nearly saturated with CO (i.e., $bP_{CO} >>$
1) at 300°C when the CO partial pressure is ~ 50 atm. As shown in
Equation 7, Y should approach $-X$ as the surface approaches CO saturation. A mechanistic basis is not obvious for the implied difference in
RDS in the two sets of experiments, but this difference is probably not
surprising in view of the great differences in experimental conditions.

The selectivity data in Table I exhibit virtually the same H_2 and CO pressure dependences as the rate data. Such a result is reasonable in terms of the mechanism shown in Figure 4. There $(CO)_s$ and H_s compete for the available surface CH_3 groups. The relative rates of production of CH_4 and C_2 chemicals should then be given by:

$$\frac{CH_4}{C_2 \text{ chemicals}} \propto \frac{\theta_{CH_3}\theta_H}{\theta_{CH_3}\theta_{CO}} \tag{8}$$

Under the assumptions developed above, θ_{CO} is approximately constant and approximately equal to the fraction of the rhodium surface not covered by carbon. Consequently, Equation 8 reduces to:

$$\frac{CH_4}{C_2 \text{ chemicals}} \propto \theta_H, \tag{9}$$

which is of the same form as that obtained for the rate of synthesis, although the constants in the two equations will differ. These constants are the terms containing the temperature dependence of the rate and the selectivity. At lower CO pressures, where $bP_{CO} \leq 1$, θ_{CO} no longer drops out of Equation 8. The selectivity ratio at low pressures should be proportional to $\theta_H\theta_{CO}^{-1}$, and hence to $P_{H_2}^{1/2}P_{CO}^{-1}$ when $bP_{CO} \ll 1$. At low pressures, then, the selectivity to C_2 chemicals should decrease with total pressure as well as with CO pressure. This effect may be sufficient to account for the lack of C_2 chemical formation at atmospheric pressure and below, although again the two sets of data available were collected under such widely differing sets of conditions and with such different analytical techniques that quantitative correlations are impossible.

To summarize the conclusions, it has been clearly shown that rhodium-based catalysts exhibit hitherto unsuspected abilities to convert synthesis gas in one step to C_2 chemicals, predominantly acetic acid and acetaldehyde, at temperatures around 300°C and pressures of 30–100 atmospheres. Addition of small quantities of manganese to these catalysts results in up to a tenfold increase in the synthesis rate, but relatively little change in the product distribution. The rate of production of methane plus C_2 chemicals by either Rh/SiO_2 or $Rh–Mn/SiO_2$ catalysts under the conditions given appears to depend on the rate of the initial addition of hydrogen to a surface carbon atom. The formation of C_2 chemicals at high but not at low pressures by these catalysts can be attributed to the greater importance of CO insertion into $Rh–CH_3$ bonds at the higher pressures. The response of selectivity to changes in H_2 and CO pressure thus provides support for the idea that $Rh–CH_3$ moieties exist on the surface and implies that these groups behave much as do their chemical analogues in solution.

Acknowledgment

This work benefited significantly from discussions with J. A. Rabo and G. A. Somorjai, who also made available results of their studies of the surface chemistry of rhodium in advance of publication. The authors also would like to express their appreciation to the management of the Research and Development Department of Union Carbide's Chemicals and Plastics Division for permission to publish this material.

Literature Cited

1. Pichler, H., Hector, A., "Kirk–Othmer Encyclopedia of Chemical Technology," 2nd ed., Vol. 4, pp. 446–489, Wiley–Interscience, New York, 1964.
2. Storch, H. H., Golumbic, N., Anderson, R. B., "The Fischer–Tropsch and Related Synthesis," Wiley, New York, 1951.
3. Fischer, F., Bahr, Th., Meusel, A., *Brennst.-Chem.* (1935) **16**, 466.
4. Schultz, J. F., Karn, F. S., Anderson, R. B., U.S. Bureau of Mines Report of Investigations **6974** (1967).
5. Vannice, M. A., *J. Catal.* (1975) **37**, 462.
6. Bhasin, M. M., O'Connor, G. L., Belgian Patent **824,822** (1975).
7. Bhasin, M. M., Belgian Patent **824,823** (1975).
8. Bhasin, M. M., Bartley, W. J., Ellgen, P. C., Wilson, T. P., *J. Catal.* (1978) **54**, 120.
9. Ellgen, P. C., Bhasin, M. M., U.S.P. **4,014,913** (1977).
10. Yao, H. C., Japar, S., Shelef, M., *Meeting of the North American Catalysis Society, 5th, 1977*, paper 18-15.
11. Vannice, M. A., *J. Catal.* (1975) **37**, 449.
12. Sexton, B. A., Somorjai, G. A., *J. Catal.* (1977) **46**, 167.
13. Wentrcek, P. R., Wood, B. J., Wise, H., *J. Catal.* (1976) **43**, 363.
14. Araki, M., Ponec, V., *J. Catal.* (1976) **44**, 439.
15. Vannice, M. A., *J. Catal.* (1975) **37**, 462.
16. Vannice, M. A., Ollis, D. F., *J. Catal.* (1975) **38**, 514.

RECEIVED June 22, 1978.

Synthesis of Fatty Acids by a Closed System Fischer–Tropsch Process

D. W. NOONER[1] and J. ORO

Departments of Biophysical Sciences and Chemistry,
University of Houston, Houston, TX 77004

*We have studied the synthesis of fatty acids by the closed
Fischer–Tropsch process, using various carbonates as promoters and meteoritic iron as catalyst. The conditions used
were D_2/CO mole ratio = 1:1, temperature = 400°C, and
time = 24–48 hr. Sodium, calcium, magnesium, potassium,
and rubidium carbonates were tested as promoters but only
potassium carbonate and rubidium carbonate produced fatty
acids. These compounds are normal saturated fatty acids
ranging from C_5 to C_{18}, showing a unimodal Gaussian distribution without predominance of odd over even carbon-num-
bered aliphatic chains. The yields in general exceed the
yields of aliphatic hydrocarbons obtained under the same
conditions. The fatty acids may be derived from aldehydes
and alcohols produced under the influence of the promoter
and subsequently oxidized to the acids.*

Fischer–Tropsch-type syntheses, i.e., catalytic reactions involving CO
and H₂ (*1, 2*), have been proposed as possible sources of organic
compounds in meteorites and on the primitive earth (*3, 4, 5, 6*). Many of
the hydrocarbons found in meteorites (*7, 8, 9, 10*) have been synthesized
in closed (*11, 12, 13*) and open-flow (*14*) systems. By adding ammonia
as a reactant, nitrogenous organic compounds including amino acids and
purine and pyrimidine bases also have been produced (*15, 16, 17*).

Fatty acids, another class of compounds found in meteorites (*18*),
are reported to be produced by Fischer–Tropsch synthesis (*1, 2, 19, 20,
21*). Consequently, we examined several of our Fischer–Tropsch products

---

[1] Current address: Spectrix Corp., 7408 Fannin, Houston, TX 77054.

0-8412-0453-5/79/33-178-159$05.00/0
© 1979 American Chemical Society

(13) for fatty acids and observed that fatty acids were produced if potassium carbonate was mixed with the meteoritic iron catalyst. Additional work (22) confirmed this observation. We have evaluated other carbonates as promoters of the Fischer–Tropsch synthesis. The results are presented in this report.

Experimental

Catalyst (0.5 g meteoritic iron) was weighed into a 4 × 20-cm glass hydrolysis tube (volume, approximately 200 mL) which served as the reaction vessel. The compound being evaluated as a promoter (0.1 or 0.3 g) was weighed into the tube and mixed with the catalyst; in some of the experiments, 5 μL of dodecanal, 1-dodecanol, or 1-pentadecene also was added. After being attached to a stainless steel Hoke valve, the reactor was evacuated for 10 min. Then at room temperature (23°–24°C), the proper pressures of deuterium and carbon monoxide were charged to give a preselected mole ratio (D_2/CO) of 1:1. The total pressure of the reactants was 2.5 atm. The charged sealed reactor was then placed in a preheated (400°C) muffle furnace. In addition to the neck (0.5 × 8 cm), upon which the Hoke valve was mounted, approximately 2.5 × 4 cm of the reaction vessel was outside the furnace.

At the end of the reaction period, the reactor was removed from the furnace, cooled to room temperature, and carefully vented to atmospheric pressure. The inside of the reactor and catalyst therein were washed with three 5-mL portions of benzene–methanol (3:1 v/v). After allowing residual solvent to drain, water soluble material in the reactor was removed with 10 mL of distilled water.

The combined benzene–methanol washings were concentrated just to dryness under nitrogen. The residue remaining was washed with three 5-mL volumes of hexane which was retained. The washed, dry residue was then dissolved in the water wash from the reactor. This alkaline water was washed with three 5-mL portions of hexane which were then combined with the aforementioned hexane wash.

To determine if fatty acids (i.e., salts of fatty acids) were present, the alkaline wash water was taken to pH 2 with hydrochloric acid and the fatty acids extracted with hexane and placed in a 20-mL vial. After removal of hexane and residual water by evaporation under nitrogen, 2 mL of 15% BF_3–methanol reagent (Applied Science Laboratories, Inc., State College, PA) was added. The vial was then sealed and held in an oil bath at ca. 75°C for 15 min. At the end of this reaction time, water was added and the methyl esters of the acids were extracted with hexane. After removal of water with anhydrous sodium sulfate, the hexane extract containing the fatty acid methyl esters was concentrated to 25–50 μL and then analyzed by gas chromatography and (in selected cases) gas chromatography–mass spectrometry.

The combined hexane washes described above were concentrated to 2–3 mL and transferred to a hexane-washed (1 × 15 cm) silica-gel column. The column was then eluted with 15 mL each of hexane, benzene, and methanol, respectively, to give hydrocarbon, aromatic hydrocarbon, and polar fractions. The fractions analyzed were evaporated to

10–200 μL. To determine if fatty acids were present, the polar fraction was evaporated to dryness in a 20-mL vial. The residue was then saponified with 2 mL of methanolic (0.5 N) sodium hydroxide by heating (oil bath) in a sealed vial for 15 min at 75°C. The remainder of the procedure was as described above for the water-soluble fatty acid salts.

The instruments used for analyses were a Varian 1200 FID gas chromatograph and a Hewlett–Packard 5730A gas chromatograph–mass spectrometer combination. The fatty acids (methyl esters) were separated using: (i) a stainless steel capillary, 154.4-m long x 0.051-cm i.d., coated with *m*-polyphenyl ether (seven-ring) fitted on Varian 1200; (ii) a stainless steel column, 3-m long x 0.4-cm i.d., packed with 10% methyl silicone on diatomaceous earth (fitted on Varian 1200); and (iii) a glass column, 1.8-m long x 0.4-cm i.d., packed with 1% methyl silicone on diatomaceous earth (fitted on Hewlett–Packard 5730A). The hydrocarbons were analyzed using columns (i) and (ii) fitted on the Varian 1200.

The Canyon Diablo (iron No. 34.6050, 10–20% nickel; American Meteorite Laboratory; *see* Figure 1) filings used as catalyst were extracted several times with benzene–methanol (3:1 v/v), and then carefully dried about 6 hr at 104°C (Canyon Diablo, nonoxidized). A portion of the Canyon Diablo filings was heated at red heat for several hours in air. After cooling, a small amount was retained for use as an oxidized catalyst (Canyon Diablo, oxidized). The remainder was then heated overnight

Figure 1.   Canyon Diablo iron No. 34.6050, American Meteorite Laboratory. The dark inclusions are nodules of graphite and troilite.

at 500°C while being slowly purged with deuterium at approximately 2 atm pressure. This catalyst was designated Canyon Diablo (oxidized–reduced).

The calcium carbonate, magnesium carbonate, potassium carbonate, potassium chloride, potassium hydroxide, sodium carbonate, and rubidium carbonate were anhydrous reagent grade materials. They were used as received.

Results

The results obtained are summarized in Table I. Chromatograms of aliphatic hydrocarbons, aromatic hydrocarbons, and fatty acids produced when potassium carbonate, rubidium carbonate, or magnesium carbonate was used as a promoter are presented in Figures 2, 3, and 4, respectively. The fatty acids obtained when a potential fatty-acid precursor (dodecanal, 1-dodecanol, or 1-pentadecene) is added to a reaction are shown in Figure 5. The mass spectogram of the methyl ester of a n-C_8 fatty acid (potassium carbonate–^{13}C used as promoter) is presented in Figure 6.

Table I. Summary of Results Obtained in Closed-System

| | | | Hydrocarbons Aliphatic | | |
| | Conditions[a] | | | | |
Run No.	Catalyst[b]	Promoter[c]	n-Alkanes (ppm)[d]	n-Alkenes (ppm)	Branched (ppm)
4-94	CDO	K_2CO_3	180	21	139
4-97	CDOR	K_2CO_3	64	86	71
4-106	CDOR	$CaCO_3$	154	—	94
4-107	CDOR	Na_2CO_3	259	—	166
4-109	CDNO	K_2CO_3	20	96	30
4-114	CDNO	KCl	72	—	26
4-120	CDNO	Rb_2CO_3	28	90	23
5-24	CDNO	K_2CO_3	44	109	25
5-27	CDNO	$MgCO_3$	43	—	23
5-32	CDNO	KOH	100	41	49
5-44′	CDNO	$K_2CO_3[^{13}C]$	18	57	20
5-58′	CDNO	K_2CO_3	—	—	—
5-60	CDNO	K_2CO_3	—	—	—
5-66	CDNO	K_2CO_3	—	—	—

[a] Temp ($\pm$ 3°C) = 400°C; time = 48 hr (except runs 5–44, 5–58, 5–60, and 5–66); D_2/CO ratio = 1:1.

[b] CDO = 0.5 g Canyon Diablo (oxidized); CDOR = 0.5 g Canyon Diablo (oxidized–reduced); CDNO = 0.5 g Canyon Diablo (nonoxidized).

[c] Runs 4–97 through 4–114, 0.3 g added; runs 4–120 through 5–44, 0.1 g added.

[d] Yields are expressed as ppm in relation to amount of CO charged ($\simeq$ 0.33 g).

[e] Dash = analyzed and nothing (or negligible amounts) detected.

Nonoxidized and oxidized–reduced Canyon Diablo meteoritic iron produced fatty acids when potassium carbonate was admixed (runs 4–97 and 4–109). The level of potassium carbonate (0.1 g vs. 0.3 g) in the catalyst (0.5 g) had no effect on the production of fatty acids (runs 4–109 and 5–24). Oxidized Canyon Diablo iron and potassium carbonate did not produce fatty acids (run 4–94), thus showing that promoter effects are catalyst dependent.

Potassium carbonate (runs 4–97, 4–109, 5–24, and 5–44) and the similar rubidium carbonate (run 4–120) promoted the synthesis of fatty acids. The other carbonates, i.e., calcium carbonate (run 4–106), sodium carbonate (run 4–107), magnesium carbonate (run 5–27), and potassium chloride (run 4–114), did not produce fatty acids. Small amounts of fatty acids were obtained when potassium hydroxide (run 5–32) was used. However, some potassium carbonate was produced in situ in this reaction.

Potassium carbonate promoted runs in which dodecanal (run 5–58), 1-dodecanol (run 5–60), or 1-pentadecene (run 5–66) was added producing fatty acids typical of those of the other potassium-carbonate-pro-

Fischer–Tropsch Synthesis of Hydrocarbons and Fatty Acids

Hydrocarbons Aliphatic	Aromatic	Fatty Acids				
Total			Normal		Branched	Total
(ppm)	(ppm)	Range	Max.	(ppm)	(ppm)	(ppm)
340	257	—ᵉ	—	—	—	—
221	1538	7:0–18:00	10:0	387	194	581
248	242	—	—	—	—	—
425	164	—	—	—	—	—
146	1080	7:0–18:0	9:0	350	193	543
98	84	—	—	—	—	—
141	531	6:0–21:0	7:0	583	285	868
178	450	6:0–21:0	9:0	497	164	661
69	<1	—	—	—	—	—
190	216	7:0–13:0	10:0	26	10	36
95	173	7:0–19:0	10:0	374	166	540
—	—	6:0–18:0	8:0	794	297	1191
—	—	6:0–16:0	8:0	356	178	534
—	—	5:0–18:0	10:0	695	278	973

ᶠ Reaction time = 24 hr; product washed from reactor with methanolic KOH.

ᵍ Reaction time for runs 5–58, 5–60, 5–66 = 44 hr. Dodecanal, 1-dodecanol, or 1-pentadecene was added. Hydrocarbon products were quantitatively similar to other potassium-carbonate-promoted reactions; yields were not calculated. Fatty-acid yields do not include the perhydro (12:0) acids; this fatty acid was recovered in amounts equal to about 1.5–2.0% of the charged alcohol or aldehyde.

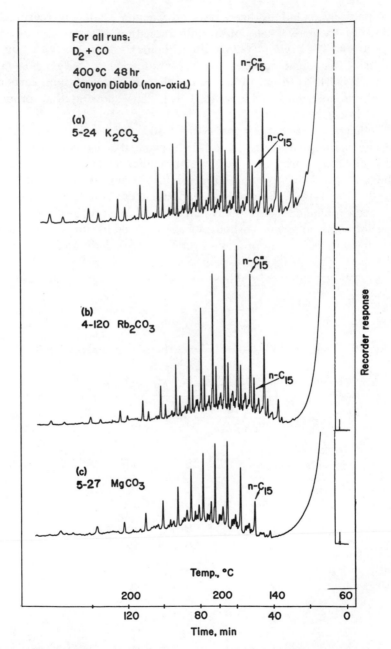

Figure 2. *Gas-chromatographic separation of perdeutero aliphatic hydrocar-*
bons in Fischer–Tropsch products. Run No. 5–24, 4–120, and 5–27.

Chromatography: Varian 1200, stainless steel capillary (152.4-m long × 0.051-cm i.d.)
coated with m-polyphenyl ether (seven ring); temperature program 60°–200°C at
2°C/min; range 10, attenuation 2.

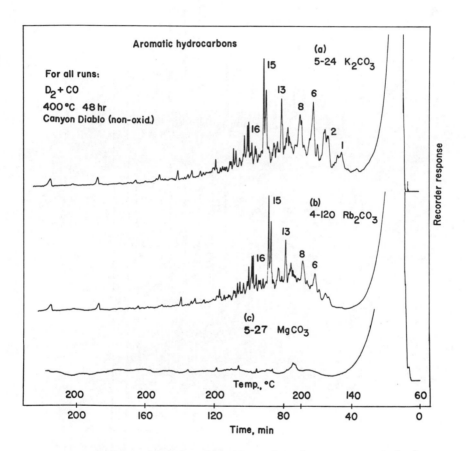

Figure 3.   Gas-chromatographic separation of perdeutero aromatic hydrocarbons in Fischer–Tropsch products. Run No. 5–24, 4–120 and 5–27.

Chromatography: same as in Figure 2. Identification of peaks: (1) methylindene, (2) naphthalene, (6) methylnaphthalene, (8) dimethylnaphthalene, (13) acenaphthene, (15) fluorene, (16) methylfluorene.

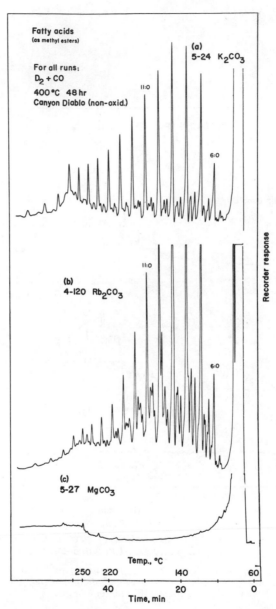

Figure 4. Gas-chromatographic separation of perdeutero fatty acids (as methyl esters) in Fischer–Tropsch products. Run No. 5–24, 4–120, and 5–27.

Chromatography: Varian 1200; stainless steel column (3-m long × 0.4-cm i.d.) packed with 10% methyl silicone on diatomaceous earth; temperature program 60°–250°C at 4°C/min; range 10, attenuation 4.

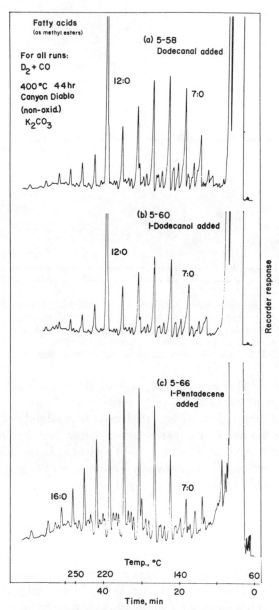

Figure 5. Gas-chromatographic separation of perdeutero fatty acids (as methyl esters) in Fischer–Tropsch products. Run No. 5–58, 5–60, and 5–66.

Chromatography: same as in Figure 4. The large peaks in 5–58 and 5–60 are not perdeutero compounds; they are attributable to oxidation of the added aldehyde or alcohol.

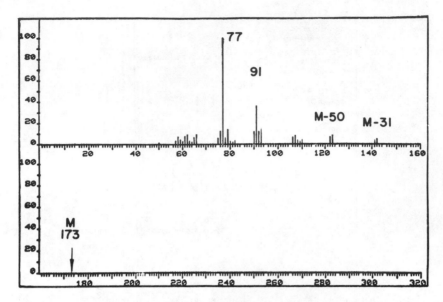

Figure 6.   Mass spectrum of perdeutero fatty acid (8:0) (as methyl ester) from run No. 5–44. The spectrum was taken as the compound was eluted from a glass column (1.8-m long × 0.4-cm i.d.) packed with 1% methyl silicone on diatomaceous earth and ionized by electron impact at 70 eV in a Hewlett–Packard 5730A gas chromatograph–mass spectrometer combination.

moted reactions. However, in the case of dodecanal and 1-dodecanol, extremely large amounts of C_{12} fatty acids were observed. No excess of fatty acid was detected when 1-pentadecene was added.

To check the possibility that a carboxylation reaction involving CO_2 from the promoter occurred, a run (5–44) was made in which potassium carbonate–^{13}C was used. As shown in Figure 6, no ^{13}C was incorporated into the fatty acids.

Discussion

The synthesis of fatty acids by a Fischer–Tropsch-type process as described in this chapter required the use of a catalyst (meteoritic iron) and a promoter. Potassium carbonate and rubidium carbonate were the only compounds evaluated which unambiguously facilitated the production of fatty acids. These catalytic combinations (meteoritic iron and potassium carbonate or rubidium carbonate) also produced substantial amounts of n-alkenes (in excess of n-alkanes) and aromatic hydrocarbons. A comprehensive study of the nonacidic oxygenated compounds produced in Fischer–Tropsch reactions (20, 21) was not made. However, in the products analyzed (all promoted by potassium carbonate), long-chain alcohols and aldehydes were detected.

The variation in promoter ability of (i) potassium and rubidium carbonates and (ii) the other carbonates may be attributable to subtle effects the active carbonates have on the surfaces and active sites of the catalyst. These effects may be caused by differences in basicity of the carbonates. According to Dry et al. (23), the promoter which is the strongest base is the most effective. The influence of base promoters on Fischer–Tropsch synthesis depends on their effect on the heat of adsorption of carbon monoxide and hydrogen on the catalyst, e.g., K_2O increases the heat of carbon monoxide adsorption at low coverage and decreases the initial heat of hydrogen adsorption.

Fatty acids (mostly branched) are produced from alkenes by addition of water and carbon monoxide (Reppe synthesis) or addition of hydrogen and carbon monoxide (Roelen or Oxo reaction) with subsequent oxidation (24, 25). However, our studies with added alkenes (e.g., Figure 5) indicated that the olefins produced were not precursors of either the normal or branched fatty acids. Oxygenated compounds, such as normal

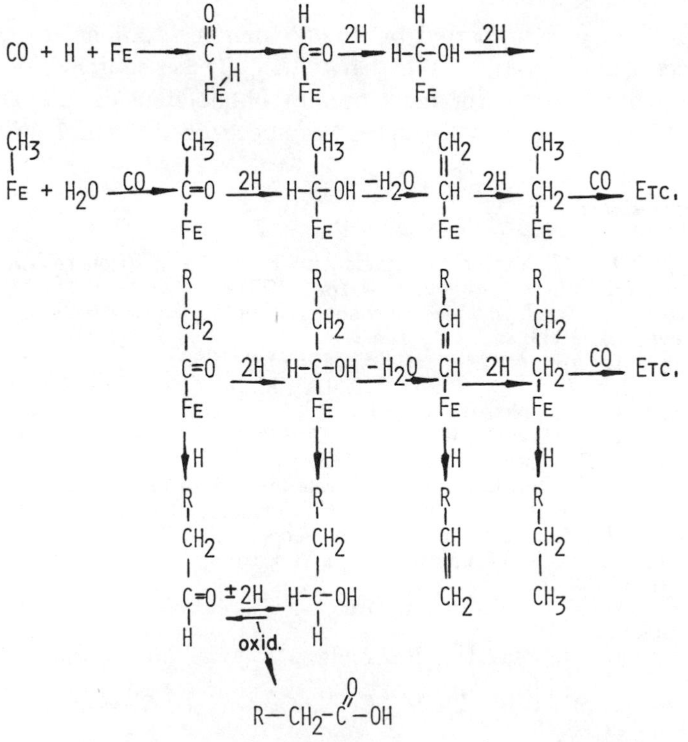

Figure 7. Scheme for the synthesis of normal compounds in K_2CO_3-promoted Fischer–Tropsch synthesis (23)

alcohols and aldehydes, could be the fatty-acid precursors since they are oxidized under the reaction conditions to fatty acids with the same C number (see Figure 5). A scheme (Figure 7) based on that of Dry et al. (23) shows how the normal compounds in the reaction products could have been formed. Oxidation could occur while the oxygenated compounds are attached to the catalyst.

Conclusion

Fatty acids in relatively high yields (usually in excess of the yields of aliphatic hydrocarbons) can be produced in a closed-system Fischer–Tropsch process using meteoritic iron as a catalyst, provided potassium carbonate or rubidium carbonate is used as a promoter. Aldehydes and alcohols or oxygenated intermediate complexes attached to the catalyst may be the source of the fatty acids.

Acknowledgment

This work was supported by grants from the National Aeronautics and Space Administration. We thank R. S. Becker and F. Feyerherm, University of Houston, for mass spectra obtained using the Hewlett–Packard 5730A gas chromatograph–mass spectrometer combination.

Literature Cited

1. Storch, N. H., Golumbic, N., Anderson, R. B., "The Fischer–Tropsch and Related Syntheses," Wiley, New York, 1951.
2. Asinger, F., "Paraffins. Chemistry and Technology," pp. 89–189, Pergamon, New York, 1968.
3. Miller, S. L., Ann, N. Y., Acad. Sci. (1957) 69, 260.
4. Oró, J., "The Origin of Prebiological Systems," S. W. Fox, Ed., pp. 137–162, Academic, New York, 1965.
5. Studier, M. H., Hayatsu, R., Anders, E., Science (1965) 149, 1455.
6. Gaucher, L. P., Chem. Technol. (1972) 2, 471.
7. Nooner, D. W., Oró, J., Geochim. Cosmochim. Acta (1967) 31, 1359.
8. Anders, E., Hayatsu, R., Studier, M. H., Science (1973) 182, 781.
9. Pering, K. L., Ponnamperuma, C., Science (1971) 173, 237.
10. Oró, J., Gibert, J., Lichtenstein, H., Wikstrom, S., Flory, D. A., Nature (1971) 230, 105.
11. Studier, M. H., Hayatsu, R., Anders, E., Geochim. Cosmochim. Acta (1968) 32, 151.
12. Studier, M. H., Hayatsu, R., Anders, E., Geochim. Cosmochim. Acta (1972) 36, 189.
13. Nooner, D. W., Gibert, J. M., Gelpi, E., Oró, J., Geochim. Cosmochim. Acta (1976) 40, 915.
14. Gelpi, E., Han, J., Nooner, D. W., Oró, J., Geochim. Cosmochim. Acta (1970) 34, 965.
15. Hayatsu, R., Studier, M. H., Oda, A., Fuse, K., Anders, E., Geochim. Cosmochim. Acta (1968) 32, 175.

16. Yang, C. C., Oró, J., "Chemical Evolution and the Origin of Life," R. Buvet, C. Ponnamperuma, Eds., pp. 155–170, North Holland, Amsterdam, 1971.
17. Hayatsu, R., Studier, M. H., Anders, E., *Geochim. Cosmochim. Acta* (1971) **35**, 939.
18. Yuen, G. U., Kvenvolden, K. A., *Nature* (1973) **246**, 301.
19. Pichler, H., Schulz, H., Kühne, D., *Brennst.–Chem.* (1968) **49**, 344.
20. Steitz, A., Jr., Barnes, D. K., *Ind. Eng. Chem.* (1953) **45**, 353.
21. Cain, D. G., Weitkamp, A. W., Bowman, N. J., *Ind. Eng. Chem.* (1953) **45**, 359.
22. Leach, W., Nooner, D. W., Oró, J., "Origin of Life," H. Noda, Ed., pp. 113–122, Japan Scientific Societies, Tokyo, 1978.
23. Dry, M. E., Shingles, T., Boshoff, L. J., Oosthuizen, G. J., *J. Catal.* (1969) **15**, 190.
24. Asinger, F., "Monoolefins, Chemistry and Technology," pp. 785–865, Pergamon, New York, 1968.
25. Falbe, J., "Carbon Monoxide in Organic Synthesis," Springer, New York, 1970.

RECEIVED June 22, 1978.

INDEX

The text of this book is set in 10 point Caledonia with two points of leading. The chapter numerals are set in 30 point Garamond; the chapter titles are set in 18 point Garamond Bold.

The cover is Joanna Book Binding blue linen.

Jacket design by Carol Conway.
Editing and production by Robin Allison.

The book was composed by Service Composition Co., Baltimore, MD, printed and bound by The Maple Press Co., York, PA.